MAPS AND THEIR MAKERS

GEOGRAPHY

Editor

PROFESSOR W. G. EAST
M.A.

Professor of Geography in the University of London

HUTCHINSON & CO. (*Publishers*) LTD
178–202 Great Portland Street, London, W.1

London Melbourne Sydney
Auckland Bombay Toronto
Johannesburg New York

First published 1953
Second, revised, edition 1962
Reprinted 1964

This book has been set in Imprint type face. It has
been printed in Great Britain by The Anchor Press,
Ltd., in Tiptree, Essex, on Smooth Wove paper.

FOR
J. C. C. CRONE

CONTENTS

LIST OF MAPS

PREFACE TO SECOND EDITION

I HAVE corrected a few errors in the original text, and have added a final chapter on contemporary cartography. The references have also been extended, and the appendices brought up to date.

Perhaps the most serious criticism of the first edition was the omission of an example of an actual map. Through the courtesy of the Royal Geographical Society and the British Council, this has been remedied to some extent by the inclusion of the map excerpt facing the title-page, details of which will be found in the list of maps.

February 1962   G.R.C.

PREFACE TO FIRST EDITION

A MAP can be considered from several aspects, as a scientific report, a historical document, a research tool, and an object of art. In this outline I have endeavoured to balance these considerations, and to regard maps as products of a number of processes and influences. Severe selection of topics has been necessary, and many of considerable interest have had to be omitted. In particular, Oriental cartography is not dealt with, nor is the cartography of individual countries. My aim has been to indicate the main stages of cartographic development to which many countries have contributed in turn.

The scope of this series does not permit extensive illustrations. I have therefore given in an appendix a list of the more important volumes of map reproductions. The best reproductions, however, are no substitute for the originals, and the only way to study them satisfactorily is by personal inspection—though this is a counsel of perfection in present conditions.

I owe a debt of gratitude to the colleagues and friends with whom at various times I have discussed much of the subject matter of this book: in particular to Prof. E. G. R. Taylor, Prof. R. Almagià, Dr. A. Z. Cortesão, and M. Marcel Destombes: to those in charge of the great map collections, particularly Mlle. M. Foncin, Conservateur des Cartes, Bibliothèque Nationale, Paris, and Mr. R. A. Skelton, Superintendent, Map Room, British Museum; and my colleagues, past and present, at the Royal Geographical Society. They have of course no responsibility for any errors or shortcomings on my part.

The outline maps were drawn by Mrs. P. S. Verity, of the R.G.S. Drawing Office.

G.R.C.

INTRODUCTION

THE purpose of a map is to express graphically the relations of points and features on the earth's surface to each other. These are determined by distance and direction. In early times 'distance' might be expressed in units of time, or lineal measures —so many hours' march or days' journey by river, and these might vary on the same map according to the nature of the country.

The other element is direction, but for the ordinary traveller, whose main concern was "Where do I go from here, and how far away is it?" the accurate representation of direction was not of primary importance. Partly for this reason, written itineraries for a long time rivalled maps, and throughout the centuries from the Roman road map to the thirteenth-century itinerary from London to Rome of Matthew Paris and even to the Underground and similar 'maps' of today, no attempt is made to show true direction. Similarly, conspicuous landmarks along a route were at first indicated by signs, realistic or conventional, and varied in size to indicate their importance, Clearly the conventions employed varied with the purpose of the map, and also from place to place, so that in studying early maps the first essential is to understand the particular convention employed.

The history of cartography is largely that of the increase in the accuracy with which these elements of distance and direction are determined and in the comprehensiveness of the map content. In this development cartography has called in other sciences to its aid. Distances were measured with increasing accuracy 'on the ground'; then it was found that by applying trigonometrical principles it was unnecessary to measure every requisite distance directly, though this method required the much more accurate measurement of a number of short lines, or bases. Similarly, instead of determining direction by observing the position of a shadow at midday, or of a constellation in the night sky, or even of a steady wind, use was made of terrestrial magnetism through the magnetic compass, and instruments were evolved which enabled horizontal angles to be measured with great accuracy. Meanwhile the astronomers showed that the earth is spherical, and that the position of any place on its surface could be expressed by its angular distances from the Equator (latitude) and a prime meridian (longitude), though for many centuries an accurate and practical method of finding longitude baffled the scientists.

The application of these astronomical conceptions, and the extension of the knowledge of the world through exploration and intercourse, encouraged attempts to map the known world—but this introduced another problem: how to map a spherical surface on a flat sheet. The mathematician came to the rescue again—with his system of projections, by which some, but not all, spatial properties of the earth's surface can be preserved on a single map. Then the astronomers discovered that the earth is not a perfect sphere, but is flattened slightly at the Poles; this introduced further refinements, such as the conception of geodetic as opposed to astronomical latitudes, into the mapping of large areas, and great lines of triangulation were run north and south across the continents to determine the true 'figure of the earth', and to provide bases for their accurate mapping.

Meanwhile, increasing demands were being made on the map maker. The traveller or the merchant ceased to be the sole user of maps. The soldier, especially after the introduction of artillery, and the problems of range, field of fire, and dead ground which it raised, demanded an accurate representation of the surface features, in place of the earlier conventional or pictorial delineation, and a solution in any degree satisfactory was not reached until the contour was invented. This again adds to the task of the surveyor who must run lines of levels and sometimes go to the extent of pegging out the contour lines on the ground. Then the archaeologist, the historian, and much later, the modern geographer had their own special requirements, and in co-operation with them the cartographer must evolve methods of mapping all kinds of 'distributions', from geological strata and dolmens, climatic regimes and plant associations, to land use and 'urban spread'. It is the present widespread recognition of the value of the map in the co-ordination and interpretation of phenomena in many sciences that has led to what may truly be called a modern renaissance of cartography.

It would be misleading to represent the stages summarily sketched above as being either continuous or consecutive. There have been periods of retrogression or stagnation, broken by others of rapid development, during which outmoded ideas have held their place beside the new. Again, cartographers have constantly realized the theoretical basis for progress, but have had to wait for technical improvement in their instruments before they could apply their new ideas. Since the easiest way to make a map is to copy an old one, and considerable capital has often been locked up in printing plates or stock, map publishers have often been resistant to new ideas. Consequently, maps must never be accepted uncritically as evidence of contemporary knowledge and technique.

In studying cartographical history, the various classes of individuals who have contributed to the map as it reaches the public must be clearly differentiated. The explorer, the topographical surveyor and the geodesist, with their counterparts on the seas, the navigator and the hydrographical surveyor, provide the data; the compiler, computer, and draughtsman work them up to the best of their ability; and finally, with printed maps and charts, the engraver or printer has his part in determining the character of the finished map. Further, in earlier centuries, when a considerable speculative element entered into ideas on the distribution and configuration of land masses, the cosmosgrapher often interpreted or applied the results obtained by explorers to fit into preconceived opinions. Professor E. G. R. Taylor has also warned the student against the booksellers and engravers, "who copied and compiled what they wanted quite uncritically, using any old maps and plates that came to hand".

Clearly the maps, many thousands in number, which have

come down to us today, are the results of much human work and thought. They constitute therefore an invaluable record for the student of man's past. It is above all this aspect with the complex problems, scientific, historical and human, which it raises, that makes the study of historical cartography so fascinating and so instructive. The appreciation of the problems of cartography and of the services which it can render to many other studies and to national development is becoming more general. Evidence of this is to be found in the establishment of the International Cartographic Association in 1961, in affiliation to the International Geographical Union. Cartographers have now a medium for the formulation and discussion of their problems and for the establishment of internationally recognised standards.

CHAPTER I

THE CLASSICAL AND
EARLY MEDIEVAL HERITAGE

IT has frequently been remarked that primitive peoples of the present day, from the Eskimo of the Canadian Arctic to the Bedawin tribesmen of the Arabian desert, have an almost instinctive ability to produce rough but quite accurate sketches on pieces of skin or in the sand, indicating the relative positions and distances of localities known to them. It may reasonably be supposed that map making began as a development of similar abilities among the early inhabitants of the Middle East and the shores of the eastern Mediterranean.

In Egypt, geometrical methods were early used for land measurement, which was stimulated by the need to re-establish boundaries after the Nile floods. These cadastral records were not, it seems, combined to make maps of considerable areas on a smaller scale, and the few 'maps' in the papyri are more in the nature of plans. The idea of maps as guides for travellers, however, was evidently current, for conventional 'maps of the nether regions' were placed in coffins for the guidance of the departed. From Assyria, there is a clay tablet with a map of part of northern Mesopotamia (c. 500 B.C.), and from Babylonia, a much later representation of the known world, shown as a circle surrounded by the sea and the heavenly bodies. Speculation such as this on the form of the Universe, and the place of the known world in it, with attempts to represent it graphically, exercised an important influence on the makers of maps.

The Greeks took over from the Babylonians, with much else of greater importance in astronomy and mathematics, the conception of the earth as a flat circular disc surrounded by the primordial ocean. In the Hellenic world the first steps in the development of scientific thought were taken by the Ionians, who were favourably placed to receive Babylonian

culture on the one hand, and on the other shared in the expanding commerce of the Mediterranean. To one of them, Anaxi-
mander, was traditionally ascribed the construction of the first
Greek map early in the sixth century B.C. The first reference
in western literature to a map occurs in Herodotus' account
of the interview between Aristagoras, tyrant of Miletus, and the
Spartans, whose help he was seeking against the Persians.
According to Herodotus, Aristagoras produced a bronze
tablet on which was inscribed "the circumference of the whole
earth, the whole sea, and all the rivers". When, however, the
Spartans learned that Susa, the Persian capital, was three
months' march from the Mediterranean coast, they refused to
listen further to him!

Our knowledge of the form and content of Greek maps,
apart from those of Ptolemy, is based upon references in the
writings of historians and geographers. From these it has been
deduced that the Greeks possessed from an early date written
itineraries and also itinerary maps of their main trading routes
in the eastern Mediterranean region. Similarly they certainly
had written descriptions of the coasts sailed by their merchant
skippers, but they do not seem to have constructed sailing
charts. As the voyages were mainly along the coasts, no doubt
written directions were preferred to charts, especially as, owing
to lack of precise instruments, the charts could not be very
accurate. There are, at least, no clear references to marine
charts. Information gleaned from sailors, however, contributed
much to the general maps, in which the coastlines necessarily
formed a considerable element.

The itinerary maps showed the stages along important
routes, for example, from the Mediterranean sea coast through
Asia Minor to the Persian capital at Susa. This was represented
as a straight line, with indications of the main features of the
country on each side of it. From sectional maps such as this, a
general map of the known world was built up. How this was
done has been shown by Sir John Myres from a study of
Herodotus. It was first sought to establish a few fundamental
lines, corresponding roughly to our parallels and meridians.
One such parallel was the royal road to Susa mentioned above;
others were provided by lists of peoples who were supposed to

succeed each other from east to west. One meridian was taken to run down the Nile, and through the Cilician Gates and Sinope to the mouth of the Ister (Danube). Since these lines were far from 'straight', the distortion introduced into the map was considerable. In this way also an east-west axis for the Mediterranean was established; since, in coasting along considerable stretches, the west coast of Italy and the south coast of France, for example, the change in direction was gradual and not easily perceptible, these portions tended to be shown as parallel to the east-west axis. The Mediterranean was thus narrowed in proportion to its length. A general principle which governed much Greek thinking then entered into the delineation of the map—namely, the symmetry of nature. Features north of the axis must be balanced by similar features to the south; the Pyrenees by the Atlas mountains, the Adriatic by the Gulf of Syrtes, Greece by the Cyrenaica promontory, and so forth. This principle was applied further afield; the Nile being thought to flow in its upper course from west to east, the unknown upper course of the Ister was made to do likewise. Emphasis on this point is necessary, for it strongly influenced later ideas on the earth's configuration. Ptolemy probably conceived his enclosed Indian Ocean as a counterpart of the Mediterranean. The frame of the world map continued to be circular, and, for the Greeks, centred at Delphi—assumptions which the philosophers often derided.

Meanwhile the progress of science was revolutionizing conceptions of the earth, and suggesting much more precise methods of fixing position on its surface. The idea that the earth was a sphere, and not a flat disc, was first advanced by philosophers of Pythagoras' school, and brought to general attention through the writings of Plato. When the spherical character of the earth was recognized, and later the obliquity of the ecliptic, astronomers were able to deduce latitudes from the proportions between the lengths of the shadow and the pointer of the sun dial. This was the forerunner of the modern method of obtaining latitude by observing the altitude of the sun at midday and applying the necessary correction from tables in the Nautical Almanac.

Thus alongside the 'mapping' of relatively small areas for

practical purposes, which corresponded approximately with what the Greeks called 'chorography', there developed very slowly the science of 'geography', that is, the mapping of the whole known world on scientific methods, what we should call cartography.

Unlike the determination of latitude, for which the axis of the earth provides an established reference datum, the problem of longitude long baffled the astronomers, for no meridian is marked out as an initial one, in the way that the Equator serves as an initial parallel. Since the earth makes one revolution in a day, more or less, it was early recognized that simultaneous observations of a celestial phenomenon such as a lunar eclipse would, through the difference in the local times at the moment of observation, give a value for the difference of longitude (1 hour=15° of longitude). Without the requisite astronomical tables or accurate portable time-keepers, the method was impracticable, though a few attempts were made to observe eclipses for this purpose. On all early maps, until the seventeenth century, longitudes were arrived at by transforming distances into their angular values in relation to the circumference of the globe. For this it was necessary to arrive at a value for the circumference, which, divided by 360, would give the length of a degree. This was done with considerable accuracy, perhaps owing to the cancelling out of errors, by the Greek astronomer Eratosthenes, who measured the meridian arc between Alexandria and Syene. The figure he arrived at for the circumference of the earth was 252,000 stadia, which assuming he employed the short stade, was the equivalent of 24,662 miles, a result only some fifty miles short of the reality. From this result it followed that a degree was equivalent to 68.5 miles. Unfortunately, this quite accurate figure was not accepted by his successors, with important results for the history of cartography.

The Greeks had also attacked the problem of the projection of the earth's surface on to a plane in order to arrive at an orderly arrangement, or graticule, of parallels and latitudes, with reference to which positions could be located. The drawing of parallels was relatively simple, at least in the restricted area for which observations were available. Eratosthenes attempted to extend two parallels eastwards on the basis of the relative directions between important places noted by travellers and also on the assumption that districts with similar climates and products would lie on or near the same parallel. In this way he established two main parallels, one running along the assumed axis of the Mediterranean (Gibral-tar-Messina-Rhodes) continued through the Taurus, the Caspian Gates and along the Imaus mountains. For another he assumed that Meroe in Egypt lay on the same parallel as south India.

The establishment of meridians, for reasons already stated, presented still greater difficulties. Without the aid of the magnetic compass, it was extremely difficult to determine the bearing of one place from another. This knowledge was derived from approximate astronomical observations, such as the position of sunrise at the equinoxes, or of the constellations in the night sky. From these Eratosthenes established an initial meridian, which assumed that the mouth of the Don, Lysi-machia on the Dardanelles, Rhodes, Alexandria, Aswan and Meroe all lay on a direct north to south line. These attempts to provide a fixed framework for the world map were criticized by his successors on the ground that the available data were insufficient. Hipparchus, the best of the Greek astronomers, corrected him in detail, and laid the foundation for further progress by compiling a table of latitudes.

With the progress of more detailed knowledge and the expansion of the known world through the achievements of Alexander the Great and the Romans, a vast mass of detail accumulated for the use of later cartographers, who could thus take up the task outlined by Eratosthenes and Hip-parchus with greater assurance of success. In the second century A.D., two names are outstanding, those of Marinus of Tyre and Claudius Ptolemy of Alexandria. The work of Marinus is known to us almost entirely from Ptolemy's refer-ences to him in his 'Geographical exposition'.[1] Marinus developed earlier ideas to construct a network of meridians and parallels, but in his world map he drew them as straight

[1]For convenience I refer to this as his 'Geography'. It was essentially a guide to drawing the world map.

lines intersecting each other at right angles. This neglect of the convergence of the meridians he considered justifiable in view of the relatively small area of the earth's surface which could be mapped and the uncertainty of much of the data. On this point he was criticized by Ptolemy, who devised two projections and also amended and supplemented Marinus' work from later information.

In discussing Ptolemy's maps it must be remembered that no manuscript older than the twelfth century A.D. has come down to us, and that it is debatable whether we have the maps as Ptolemy drew them or indeed whether he actually drew maps at all. Apart from general sections on cartography, projections, and Marinus' ideas, the 'Geography' is essentially an extensive table of the geographical co-ordinates of some 8,000 localities. Since there were very few actual astronomical observations available, he obtained the positions of these localities by a careful study of itineraries, sailing directions, and topographical descriptions of various countries. In this he endeavoured to allow for the windings of routes, by reducing many of the itinerary distances, for he shared that distrust of travellers' estimates which Marinus displayed when he wrote that "merchants, being wholly intent on business, care little for exploration, and often through boasting exaggerate distances". The simplest method of arriving at the co-ordinates would be to construct maps from such data, and then to read them off from the network of meridians and parallels. It is also hard to believe that having undertaken all this laborious preliminary work he would have refrained from supplementing his text with maps. This is not to say that we have the maps as they left his hands. Of the world map it is definitely stated that this was drawn by one Agathodaimon of Alexandria, who may have been contemporary with Ptolemy.

There are also inconsistencies in the text, and between text and maps. Father Joseph Fischer, the great student of the 'Geography', considered that the maps were originally drawn by Ptolemy, but that they became separated from the text, and that both underwent modifications before they were brought together again. A recent student, Leo Bagrow, however, has put forward a more drastic interpretation. From a critical study of the text, which admittedly lacks unity, he believes that it was put together from Ptolemy's writings by a Byzantine scribe in the tenth or eleventh century, and from the tribal names in European Sarmathia (western Russia), he concludes that the maps were not drawn until the thirteenth century. He has further found a record that a Byzantine, Maximos Planudes (c. 1260–1310) who possessed a manuscript of the text, drew a set of maps for it, from which Bagrow believes the later MS. maps stem. Though it is clear that the maps as they have survived are not the unaltered work of Ptolemy, it does not necessarily follow that he did not draw maps (the case of Agathodaimon and the world map suggest that quite early his data were being used for maps). What is of more importance, through the Ptolemy manuscripts, whatever is the truth about their history, there was transmitted to Renaissance scholars, a vast amount of topographical detail, which profoundly influenced their conception of the world.

The manuscript maps fall into two classes, one consisting of the world map and twenty-six regional maps. It was this set which accompanied the Latin translations of the fifteenth century and were used for the earliest printed editions. The second class contained sixty-seven maps of smaller areas. The world map is drawn on the more elementary of the two projections described by Ptolemy—a simple conic with one standard parallel. The special maps are on a rectangular projection with straight parallels and meridians intersecting at right angles; they indicate the boundaries of provinces, and the relative positions of important nations, as well as cities, rivers, and mountains.

It is necessary to dwell on his maps in a little detail on account of their influence upon the renaissance of cartography. From the second until the early fifteenth century, they were almost entirely without influence on Western cartography: they were, however, known to the Arab geographers, who possessed translations of his works, and through them seem to have had some influence on fourteenth-century cartographers such as Marino Sanudo. With the translation of the text into Latin in the early fifteenth century, Ptolemy dominated European cartography for a century, and, through his insistence upon the scientific basis, his influence promoted cartographical progress. On the other hand, in several ways his ideas hindered the development of an accurate map of the world. One of his principal mistakes was the adoption of a value for the length of a degree equivalent to 56½ miles, in contrast to Eratosthenes' more accurate value. Thus when transforming distances into degrees, he obtained greatly exaggerated figures, an exaggeration intensified by the common tendency of travellers to overestimate distances covered. The longitudinal extent of the Mediterranean, for example, he fixed at 62°, instead of 42°, and in the same way he exaggerated the easterly extension of Asia, placing its eastern shores some 50° too far east. This was however a reduction of 45° on the figure adopted by Marinus. He also incorporated erroneous conceptions on the configuration of the old world: for example, he greatly overestimated the size of Taprobana (Ceylon) and overlooked the peninsular form of the Indian sub-continent, or perhaps confused it with Ceylon; he conceived the Indian Ocean to be a land-locked sea, extending the south-eastern African coast-line eastwards to meet a southerly extension of what he probably intended to represent the Malay peninsula. Another conspicuous error is the easterly direction he gave to Scotland, probably due to a mistake in joining two sectional maps together. His representation of the hydrography of northern Africa, which showed a great eastward-flowing river, perhaps the Niger, ending in a central swamp, was sometimes followed until the early nineteenth century. Less erroneous was his delineation of the Nile, rising from lakes at the foot of the Mountains of the Moon, some degrees north of the Equator. It is useful to keep these misrepresentations in mind when studying the maps of the Renaissance and to note their gradual elimination as exploration progressed.

The Romans seem to have been singularly unconcerned with Greek achievements in scientific cartography. For them a map remained a practical aid to the journeys of their officials and the campaigns of their legions. If we were to judge from the sole surviving example of any size, we would conclude that many were little more than graphical renderings of written itineraries. This example, and in a very late copy at that, is

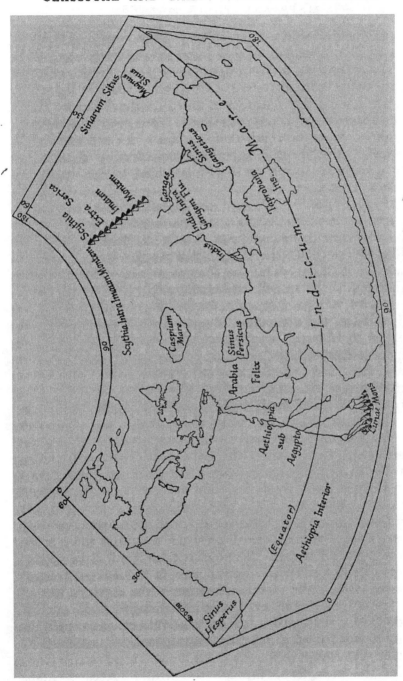

Ptolemy's World Outline, from the Rome edition of 1490

the 'Peutinger Table', so called after the sixteenth-century humanist who once owned it. The Table is essentially a road map of the Roman Empire, constructed to fit a long, narrow roll, no doubt so that it might be carried conveniently. The roads are indicated by straight lines, and distances are marked between each stage. Changes in direction are shown by 'kinks', and branch roads diverge similarly. Thus true directions are neglected, with the result that the shape of countries and the relative positions of features are considerably distorted. It was simply, as it was no doubt intended to be, an efficient guide to road users.

From references in literature, describing the use of maps in campaigns, and their value to commanders, it is clear that all Roman maps cannot have closely resembled the Table. Some idea of their general character may be formed from the references to the most famous Roman map, the *Orbis terrarum*, or 'survey of the world', executed by M. Vipsanius Agrippa, son-in-law of the Emperor Augustus who authorized the undertaking, and supervised its completion after the death of Agrippa in 12 B.C. Pliny bears testimony to Agrippa's 'extra-ordinary diligence', and the care he bestowed upon the work, which was placed in the Porticus Vipsania at Rome for the information of the citizens. In his topographical descriptions of countries contained in his 'Natural History', Pliny, who had seen the map displayed, several times quotes Agrippa on the dimensions and boundaries of countries, presumably obtained from the map. Since these quotations refer to seas, rivers, mountains, islands, provinces, and towns, it was drawn in great detail. The basis of the map was no doubt the distances along the Roman road system and official returns supplied by the provincial administrators. Varied opinions have been expressed on the probable shape of the map, but the majority hold that it was circular. The popularity in late Roman times of the small T-O maps mentioned later is indirect evidence of this. In view of the official character of the Agrippa map, it was doubtless circulated in copies on a reduced scale, such for example as the map which Eumenius relates was studied by schoolboys at Autun in the fourth century. A case can be made out for the persistence through the Middle Ages of maps deriving ultimately from the Agrippa model, of which the Hereford *mappa mundi* is an example.

The contrast often drawn between 'practical' Roman and 'scientific' Greek cartography tends to be exaggerated. While it is true that Greeks had arrived at a more scientific conception of the essentials, methods of obtaining the requisite data lagged behind theory. It was only at the end of the period that Greek cartography culminated in the work of Claudius Ptolemy, and even then it had serious limitations. It is not difficult to believe that to the Romans a map based upon the foundation of the road system was more acceptable than the work of the Greek geographers, however scientifically conceived.

Space does not permit the examination in great detail of early medieval cartography: but certain points should be kept in mind. For several centuries, geographical knowledge was at a standstill, if not in retreat. Consequently geography and, to a greater extent, cartography became merely a routine copying of the accepted authorities, into which an increasing number of errors were introduced. Many of the so-called maps of this period were reduced and simplified diagrams, inserted in standard descriptions of the known world. A common type are the numerous so-called T-O maps, in which, oriented with the east at the head, the O represented the boundary of the known world, the horizontal stroke of the inset T the approximate meridian running from the Don to the Nile, and the perpendicular stroke the axis of the Mediterranean. Other versions of this occur in a rectangular frame, which may have been adopted as more economical of space, or as complying with Biblical references to the 'four corners of the earth'.

The main type of circular world map, or *mappa mundi*, which was perpetuated through this period appears with little doubt to be related, though distantly, to the world map of Agrippa, modified to bring it into conformity with orthodox Christian theology. Here again there are some variations in shape, for example, the map of Henry of Mainz in Corpus Christi College Library, Cambridge, is elliptical; this shape may have been adopted to fit more conveniently on the page of the manuscript. In any event, the content of such maps does not vary significantly from that of the circular type.

The largest and most interesting surviving example of the circular world map is the *mappa mundi* now preserved in Hereford Cathedral. Though dating from as late as *circa* A.D. 1300, it is clearly but the last in a long chain of copies. One of the links is the Hieronymus map of about A.D. 1150, a portion of a world map now in the British Museum. There are several reasons for holding that it derives from a Roman original, apart from inscriptions upon it which associate it with Orosius, the fourth-century writer, and refer to the survey of the world by 'King' Agrippa. Broadly, the area it represents corresponds to the limits of the Roman Empire with an extension to include the conquests of Alexander the Great. The provincial boundaries shown correspond fairly closely to those of the time of Diocletian. The shapes given to certain countries resemble those assigned to them in popular writers of Roman times, and some groups of the named towns, though jumbled up on the map, correspond to sections of the Antonine Itinerary. Accepting this Roman pedigree for the Hereford map, we must allow that it had undergone significant alterations at the hands of Christian theologians. The centre of the map is Jerusalem, not a very serious distortion, since the centre of the original may well have been in the vicinity of Rhodes. It is disputed whether the Roman map was oriented with the east at the head, but this would not be a difficult alteration, and it enabled the Christian scribe to show the terrestrial paradise in a position of honour. Again, the area of Palestine has been considerably enlarged, as one of the objects was to depict sites hallowed in Holy Scripture.

The general scheme resembles that of the T-O maps, though somewhat distorted by the emphasis placed on Palestine, Asia Minor, etc. Rome, Antioch, and Paris are drawn very conspicuously, the prominence of the latter suggesting that one of the 'links' had been the work of a French scribe. Other cities and towns are represented by conventional drawings of towers and gates; mountains and rivers are numerous, the former in a conventionalized profile. Most of the space, which would otherwise be empty, is filled with neatly executed drawings depicting themes from the popular histories and bestiaries of the time. Indeed the whole is as much an encyclopedia of medieval lore as a map, and provides fascinating material for study.

Of greater interest for this outline is the fact that, though mainly copied from older sources, it contains additions showing that interest in cartography had not entirely died out. Several towns prominent in the English administration of Gascony in the thirteenth century have been inserted, and there are traces of a commercial route from north Germany towards the Rhine from an earlier date. The depiction of the British Isles on the Hereford map, though crude, is later than the general content, with medieval forms of town names, and four cities in Ireland. The representation of the Trent-Ouse river systems of northern England also indicates local knowledge. The evidence for medieval cartographic activity in Britain is not however confined to this map. Dating from A.D. 1250 are the four maps of Matthew Paris, the St. Albans chronicler: in one case based on a straight line itinerary, from Dover to Newcastle. Though presenting difficulties in interpretation, they nevertheless show that attempts to draw maps, however crude, were being made. Much more striking is the 'Gough' map of the following century (circa 1360) with its elaborate road system and careful distinctions in the status of the towns depicted. It has been suggested by R. A. Pelham that this may be a copy of an official road map prepared for the use of Edward I.

As has been remarked above, Ptolemy's 'Geography' was almost entirely without influence on the medieval West, but it was known and studied in Byzantium, to whose scholars it is possible we owe the extant maps. Further study of the precise role of Byzantine culture in the history of cartography may produce important results. There was also another centre where Ptolemy's influence exerted itself, namely the Arab world. The text of the Geography was translated into Arabic in the ninth century, and versions of his maps were known to Arab scholars, for example, Mas'udi, in the following century. Except in one instance, however, there was no direct contact between Arab and European cartography. In the twelfth century, the geographer el-Idrisi was welcomed at the court of Roger II, the Norman king of Sicily, and there compiled a world map, with a written description, incorporating Arabic and Western sources, the latter being obtained for him by royal order. It is generally assumed that they were written reports by sailors and merchants. But his description, for example, of the English coasts has some striking resemblances to the outline of the earliest marine charts, though the names do not correspond. As will be shown in the next chapter, these are considered to have originated about A.D. 1250. Were these charts based on material similar to Idrisi's, or should their origin be set back a century? This is another problem which would well repay investigation.

If the direct influence of Arab cartography on western Europe was not great, works on astronomy and mathematics translated from the Arabic, as will be seen later, stimulated progress from the thirteenth century onwards.

It is clear that by A.D. 1300, apart altogether from the great advance in sea charts discussed in the next chapter, cartography was beginning to emerge from its 'dark ages'. There is certainly no clean break at this date, for features of the medieval *mappae mundi* persisted for long in Renaissance maps. But widening horizons presented greater incentives to the cartographer, and spurred him to the solution of more complex problems than had faced his medieval predecessor, confined by poor communications to western Europe, threatened on almost all sides, and dependent on the restricted resources of monastic libraries.

REFERENCES

BAGROW, L., The origin of Ptolemy's 'Geographia'. (*Geografiska Annaler* 27, 1945, 318–87.)

BEVAN, W. L. and PHILLOTT, H. W., Medieval geography; . . . the Hereford Mappa Mundi, 1873.

FISCHER, J., Claudii Ptolemaei Geographiae Urbinas Graecus 82 (with commentary). 4 vols., 1932.

KLOTZ, A., Die geographischen Commentarii des Agrippa und ihre Über-reste. (*Klio* 24, 1931, 38–58, 386–470.)

MILLER, K., Mappae mundi; die ältesten Weltkarten. 6 pts. Stuttgart, 1895–98.

—— Itineraria Romana, Stuttgart, 1916.

MYRES, J. L., An attempt to reconstruct the maps used by Herodotus. (*Geogr. Journ.*, 8, 1896, 605–31.)

THOMSON, J. O., History of ancient geography, 1948.

UHDEN, R., Zur Herkunft und Systematik der Mittelalterlichen Welt-karten. (*Geogr. Zeits.*, 37, 1931, 320–40.)

CHAPTER II

THE EVOLUTION OF
THE MEDIEVAL SEA CHART

TOWARDS the end of the thirteenth century there came into
use in western Europe a type of chart which was a great
advance upon any other product of medieval cartography so
far considered. In their essentials these charts marked a complete break with tradition: their fundamental feature was that
they were based on direct observation by means of a new instrument, the mariner's compass. On them the coasts of the Black
Sea, the Mediterranean, and south-west Europe were laid down
with considerable accuracy, and the outline thus established
was followed by map makers for several centuries, without sub-
stantial modification until the fixing of position by astronomical
observation became general in the eighteenth century.

These charts are frequently referred to as portolans, but
as the term *portolano* is properly applied only to written sailing
directions, this usage is confusing. Some would call them simply
medieval sea charts without distinction; but, to indicate the
particular type characteristic of the fourteenth and fifteenth
centuries, it is convenient to compromise with the term
'portolan chart'.

These charts have survived as single examples or as 'atlases'.
The latter are in the main merely the standard chart divided
into sections, sometimes bound with a calendar, world map,
or astronomical data. The total number of charts surviving
from the fourteenth century is not great: probably not many
more than a score, and only seven draughtsmen can be identified with certainty; of these three, Petrus Vesconte, Angellino
de Dalorto, and Johannes de Carignano, worked at Genoa;
two, Perrinus Vesconte and Francesco Pizigano, at Venice;
and two, Angellino Dulcert[1] and Guillelmus Soleri at Majorca.

[1]Dulcert is probably the Catalan rendering of Dalorto, and these two
names are generally assumed to refer to the same cartographer.

29

The earlier charts, it should be noted, are by Italian draughts-men. In addition there are three 'world' maps closely related to these charts, one certainly and two probably dating from this century.

The best known is the great Catalan atlas of 1375, now in the Bibliothèque Nationale, Paris, and attributed to Cresques the Jew.[1] They are all by Majorcans or have legends partly in Catalan, so that it would appear that as the century pro-gressed, the centre of cartographic activity was definitely established at Majorca.

The charts are drawn on single skins of parchment, which generally preserve their natural outline, and they range in size from 36×18 inches to 56×30 inches. The coastline is in black, often faded and faint, but with its outline emphasized by the long series of names of ports and coastal features, written perpendicularly to it. These names are in black, save for those of important harbours which are in red. Small islands, including river deltas, are in solid colour, red or gold, and rocks or shoals are indicated by small crosses or dots in black or red. In what is sometimes referred to as the 'normal portolan', there is little detail inland; occasionally a few rivers, mountain ranges, and vignettes of larger cities with banners. These are often carefully drawn and brilliantly coloured. The whole effect, especially on the later and more elaborate examples, is extremely decorative. It must be remembered that charts, being the working tools of a hazardous profession, were liable to be lost or discarded when worn out, so that it is not surprising that few have survived. The existing decorative charts were no doubt prepared for wealthy shipowners or merchants, and, preserved in their libraries, escaped the hazards of the sea. There is no reason to doubt, however, that they resemble in essentials those in everyday use.

The charts to which particular reference is made in the following paragraphs are : (i) the 'Carte Pisane', so called because it once belonged to a family of Pisa. It is probably of Genoese origin. The draughtsman is not named, and it is undated, but generally assigned to the late thirteenth century. It extends from the Black Sea to southern England (extremely crudely

[1] See p. 40.

drawn) and the neighbouring coasts of western Europe. The eight winds are named, and there is a scale. Two series of lines radiate from centres of circles near Smyrna and west of Sardinia. Areas outside the circles are enclosed in rectangles, and divided into squares, by which doubtless sectional charts were incorporated into the whole. England is not treated in this fashion. The names, in red and black, are confined almost entirely to the coasts, and there is little interior geography. (ii) Atlas of Petrus Vesconte, 1318. This is simply divided into nine sections, covering approximately the same area as (i), but more carefully drawn, with a somewhat better outline of southern England. (iii) Chart by Perrinus Vesconte, 1327. Its general features resemble those of (ii), but the coastline of southern England is much improved. A few vignettes of inland cities have been added. (iv) Chart by Angellino de Dalorto, c. 1325, very carefully drawn and finely coloured. It extends from the Black Sea to the Baltic, but the outline of the latter is extremely poor, and not based on a detailed survey. To the characteristics of earlier charts have been added the Rhine, Elbe, Danube and other rivers; mountain ranges in green, and many cities. Owing to the severe contraction of the northerly extent of Europe, the interior physical features are greatly distorted. It is plainly a stage in the transition from a purely marine chart to a world map.

All these charts have scales, with the main divisions subdivided into fifths by dots. Curiously enough, the unit of length is never stated. As a result of a number of measurements, Professor Wagner concluded that two units were employed; in the eastern Mediterranean, a mile of about 4,100 feet, or two-thirds of a modern nautical mile. For the Atlantic coast he obtained a value of 5,000 feet. As the result of this discrepancy in scale, the Atlantic coastline is conspicuously restricted in extent.

These charts have several features in common. The area they cover comprises the Mediterranean and Black Seas with a portion of the Atlantic coasts of Europe. South of the Strait of Gibraltar, the charted coast extends a short distance beyond the termination of the Atlas Mountains: to the north, the coasts of Spain, France, southern England and the Low

Countries are depicted less accurately. Beyond, the outline
becomes much less precise, and those charts which attempt
to show the Baltic do so very sketchily, in striking contrast
to their accuracy elsewhere. It is significant that the portions
which depict the coastlines most correctly correspond in
general with the regions with which Genoese and Venetian
trading activities were highly developed. Venice dominated
the trade of the Black Sea, where a factory had been established
at Tana on the Sea of Azov in the twelfth century. The Genoese,
their keen rivals, were firmly entrenched in the eastern Mediterranean, and at this period, following their victory over
Venice in 1298, were at the height of their prosperity. Both
city-states were also established in the ports of northern
Africa, and their fleets were ranging as far as the Low
Countries.

The second feature which immediately attracts attention is
the system of lines with which they are covered. From two
points, in the western and eastern Mediterranean, sixteen or
thirty-two lines radiate over the chart, and on the circumference
of circles about these points similar subsidiary centres are
equally spaced, so that the whole chart is systematically
covered. On later charts, these lines spring from the centres of
'compass roses', and their purpose is clearly to represent lines
of direction.

On the earlier charts, these groups of radiating lines are not
directly associated with a compass, or wind 'rose'. The cardinal
points are shown towards the margins of the charts, sometimes
by their names alone, in other instances by various symbols,
such as drawings of heads. On Petrus Vesconte's chart of 1311,
a cross within a circle, also containing the scale, may be intended
to indicate the four cardinal points. Angellino de Dalorto's
chart of 1325 represents a further stage; on it the north is
marked by a circle containing an eight-point star, which it is
reasonable to suppose indicates the principal points. It is not
until the Catalan map of 1375 that a complete compass rose is
found forming an integral part of the system of radiating lines.
Since the arrangement of lines is much the same on all, it may
be assumed that those on the earlier charts were also intended
to represent compass bearings.

If these charts are compared with modern examples the central axis of the Mediterranean is seen to be rotated about 10° to the left. It is believed that the magnetic variation in the Mediterranean at this period was approximately ten degrees east, and this suggests that the chart was drawn to bring the line indicating magnetic north into the vertical.

No contemporary explanation of this system of direction lines has survived, but since navigators were a conservative race, we may take that given by John Rotz, a sixteenth-century expert, in his 'Boke of Hydrographie'. His explanation is somewhat complicated, but in essence he instructs the navigator how by the aid of a pair of dividers he can find the line or ray most nearly parallel to the course between any two points on the chart, and then read the correct bearing from the nearest compass rose. This was later done more easily with a parallel ruler. To prevent mistakes the lines were generally drawn in alternating colours. The purpose of this elaborate system of 'roses' and radiating lines was therefore to enable a course to be determined rapidly by providing a number of reference points distributed over the chart. It thus became possible to plot a course over a considerable extent of sea, in contrast to the coastwise navigation with the aid of the details in the written *portolanos*. This was the fundamental difference between the chart and the book of sailing directions.

It will be noticed that none of these charts is provided with a net of parallels and meridians. In their construction, no account was taken of the sphericity of the earth, the area covered being treated as a plane surface, and the convergence of the meridians neglected. The consequences of this were not serious, owing to the small range in latitude involved. The direction lines therefore approximate to loxodromes (lines of constant bearing).

It was not until the early sixteenth century that charts were provided with a scale of latitudes. As long as European navigators were working mainly in enclosed waters, and sailing coastwise from point to point, they had little need to observe latitudes, and in fact even in the seventeenth century it was not customary for navigators in the Mediterranean to take such observations. When maritime activity passed out of these

B

confined waters and spread across the great oceans, observations for latitude served as a check on dead reckoning. As a further consequence, the sphericity of the earth's surface could no longer be overlooked, and the problem of choosing a projection which would allow a line of constant bearing to be represented on the chart by a straight line became acute. The solution finally achieved is the projection which bears the name of Mercator.

For the reasons set out above, it seems clear that the portolan charts were from the beginning closely related to the compass, and that it was the introduction of this instrument which made their construction possible. Some students, however, have held contrary views. Professor Wagner, in his study of the scales, equated the shorter mile used in the Mediterranean to an ancient Italian-Greek unit, and that used on the Atlantic coasts with the later Roman mile; from this he argued that the Mediterranean portion of the chart must date back to a period long before the introduction of the compass. Though no chart has survived from Roman or Greek times, a few books, or portions of books, of sailing directions are extant, and it is probable that details from these were incorporated in the portolan charts.

As for the *portolano* or book of sailing directions, known later in England as a 'rutter of the sea', it is true that there is evidence for its existence in medieval times, before the introduction of the chart. The oldest known example is contained in Adam of Bremen's 'Ecclesiastical History', written in the twelfth century. This has the appearance of being a very summary version of a more detailed document. In a few lines it sets out the stages on a voyage from the mouth of the river Maas to Acre in Palestine. It gives the distances from point to point by the number of days' sailing, with an approximate indication of the direction to be followed. The only point mentioned on the English coast is 'Pral', perhaps Prawle Point, or Portland Bill. Unless much more detailed instructions existed, it is difficult to imagine the portolan charts being produced from material of this type. At the very least, the use of the compass would simplify the work greatly and give much improved results. Others have argued that the method of indicating direction by lines radiating from a centre was of ancient date, and is known to have been occasionally used in medieval times. In the earlier instance, however, the division of the circle was into twelfths and not into eighths, as on the compass rose. The best answer to these objections lies in a closer examination of the purpose of this system of direction lines, which has been discussed above—for it provided a method of determining a course which could not be found from a written *portolano*, and this method depended upon the use of the compass.

In attempting to determine the date of the first appearance of these charts, therefore, we can be guided to some extent by what is known of the history of the compass. A primitive form was probably in use in the twelfth century, consisting of a needle thrust through a piece of wood and floating in a bowl of water. An improved type appeared about the year 1250, in which the water was dispensed with, the needle being balanced on a pin. With the addition somewhat later of the compass card, the compass as we know it today had in essence evolved. This development may well have been decisive in the evolution of the portolan chart. By means of the card, bearings could be taken with comparative ease and rapidity. Keeping this in mind, we may turn to consider the other evidence for dating. The 'Carte Pisane' was probably drawn towards the end of the thirteenth century, and the first dated chart is from the year 1311. Since their form had then become stereotyped, the original can scarcely be later than the third quarter of the thirteenth century, and it is about that time that the earliest references in literary sources to sea charts are found.

There is an incident on record which shows that charts were in use in A.D. 1270. In that year, King Louis IX embarked on the Mediterranean for his crusade in North Africa. Soon after setting sail, the fleet was dispersed by a prolonged storm. When it had passed, the King was anxious to know the position of his ship, and the pilots were able to point out on a chart that they were approaching Cagliari. Writing at about the same period, Raymond Lull includes the chart among the instruments employed by seamen. We may therefore assume that the portolan chart originated in the period 1250–75.

Some time seems to have elapsed before they were generally adopted by the navigators of the Mediterranean. As late as 1354, King Peter of Aragon considered it necessary to issue an ordinance requiring each war galley to be furnished with two sailing charts, an action, incidentally, which must have stimulated the output of the Catalan chart makers, who were then taking the lead in advancing cartography.

If the seamen were not quick to adopt the chart, some scholars appreciated its value at an early date. When Marino Sanudo drew up his appeal to the Pope to revive the crusading movement against the Turk, embodied in his 'Liber Secretorum fidelium crucis', he included not only a written *portolano*, but a set of charts, drawn by Petrus Vesconte, to provide accurate information for the planning of the proposed campaign. Vesconte was obliged to combine new and old sources of information as best he could—as may be plainly seen in his chart of the eastern Mediterranean, where the respective contributions of portolan chart and medieval map can be distinguished at a glance. A similar preliminary attempt to fit the new chart into the traditional framework can also be observed in Dalorto's work—for he shows himself well acquainted with the medieval world maps. He inserts in the margin a small T-O map; he marks the limits of Europe with stereotyped phrases ('Europa incipit ad Gallicia', 'Finis Europae'), and has vignettes, e.g. of the Tower of Babel, which recall the designs of the Hereford Map. He was therefore a scholar, rather than a chart maker, but was in touch with contemporary progress.

From successive charts of the British Isles, we can form an idea of the time required for new surveys to be incorporated in the standard chart. On the 'Carte Pisane', Britain is represented in a very crude form, and lies outside the framework of the main chart. From about 1325 a complete representation of the British Isles is attempted, but obviously the cartographer had little information about Scotland, and though he was better informed about Ireland, that island is made too large in comparison with England. An examination of Perrinus Vesconte's outline of 1327 shows that in fact the only area known at all accurately to him was southern England from the Bristol Channel to the Thames estuary. To the north, the only conspicuous feature on the east coast is the Humber, and even the projection of East Anglia is omitted. The west coast is even less adequately delineated; the peninsula of North Wales is lacking, and the coast beyond is represented by a large semi-circular bay. As southern England is much too small compared with the rest of the country, it is plain that a piece of relatively accurate survey has been fitted to an older highly generalized outline of the whole island. Scotland appears almost severed from England by two rivers, a feature which recalls the Matthew Paris maps.

The source of this knowledge of southern England is to be sought in the development of commercial relations between the city states of north Italy and western Europe. The annual fleet, known as the 'Flanders galleys', which sailed from Venice to the Low Countries, is first mentioned in 1317. Part of this fleet traded at Southampton, Sandwich, and London. It is said that Vesconte was consulted by the Venetian authorities when the fleet was being organized. He would therefore be well placed to obtain from the commanders of the galleys on their return the results of their observations along the English coast. This would account for the improvement shown by the chart of 1327 in comparison with the 'Carte Pisane'. Throughout the century, no essential change was made in this outline; the variants appear to show deterioration from the original model, rather than fresh sources of information. At the beginning of the fifteenth century, however, the outline of the southern coast was improved, as may be seen on the chart of G. Pasqualini, Venice, 1408. If the above hypothesis is correct, it tends to show that a considerable stretch of coast could be surveyed and incorporated in charts within a few years. The charting of the Mediterranean might therefore have been the result of surveys carried out in a relatively short period of time, without much reliance upon the work of preceding centuries. Charles de la Roncière suggested that the Genoese admiral, Benedetto Zaccaria, had the opportunity of initiating and supervising such a survey, for he successively commanded the Byzantine, Genoese, Castilian, and French fleets. In the latter capacity he was actually in charge of French naval operations against England in 1298. There is however no direct evidence to connect him with the development of the portolan chart.

To sum up; the available evidence suggests that these charts were introduced in the second half of the thirteenth century, that they were based upon the use of the mariner's compass, and that the navigators and cartographers of northern Italy, especially of Genoa and Venice, played a predominant part in their development. Their history is a good instance of the response of technicians to a new social demand, in this case the need of the commercial communities of Italy to develop communications with their expanding markets. The achievement of these thirteenth-century cartographers was a notable contribution to knowledge, and one which was not surpassed for centuries.

REFERENCES

ANDREWS, M. C., The British Isles in the nautical charts of the XIVth and XVth centuries. (*Geogr. Journ.*, 68, 1926, 474–81.)

BEAZLEY, C. R., The dawn of modern geography, vol. 3, 1906.

CARACI, G., Italiani e Catalani nella primitiva cartografia nautica medievale, Rome, 1959.

HINKS, A. R., ed., The portolan chart of Angellino de Dalorto, 1325. R. Geogr. Soc., 1929.

KRETSCHMER, K., Die italianischen Portolane des Mittelalters. (*Veröffentl. Inst. f. Meeresk., Hf.* 13, *Berlin*, 1909.)

MOTZO, B. R., Il compasso da navigare. Cagliari, 1947.

TAYLOR, Eva G. R., The haven-finding art, London, 1956.

—— Mathematics and the navigator in the thirteenth century. (*Inst. of Navigation*, London, 1960.)

WAGNER, H., The origin of the medieval Italian nautical charts. (*Rept. 6th Internat. Geogr. Congress*), London, 1896.

WINTER, H., The true position of H. Wagner in the controversy of the compass chart. (*Imago Mundi*, 5, 1938, 21–6.)

CHAPTER III

CATALAN WORLD MAPS

ANOTHER notable stage was reached in the fourteenth century, when European cartographers made the first attempt since classical times to include the continent of Asia within their world picture on the basis of contemporary knowledge. The results of these efforts are embodied in the series of Catalan world maps.

In the first half of the fourteenth century the Catalan school, largely Majorcans, took over from the north Italians the lead in cartographic progress, though rather as successors than as innovators. During the preceding century the Majorcans had earned a great reputation among the peoples of the western Mediterranean for their maritime prowess. After their incorporation in the Aragonese confederation (1229), the three ports of Palma, Barcelona, and Valentia formed the basis of a commercial enterprise which extended to most of the north African ports as far as Egypt, and beyond to Syria. Early in the century, the population had been augmented by Jewish refugees from the Almohed persecutions, and this element strengthened commercial relations especially with Morocco. Trade was also stimulated by the aggressive policy of the able rulers of Aragon, and diplomatic agents appear by 1300 to have reached as far afield as Persia. But these Jewish refugees also included scholars who could interpret the works of Arab scientists, and this contact between practical and skilled seamen and those versed in cosmography and astronomy was fruitful. These sciences were also encouraged by the enlightened House of Aragon, under whose patronage Barcelona became a centre for the diffusion of Arabic knowledge, and therefore of advance in mathematics, astronomy and the construction of instruments.

This intellectual ferment was not without its influence on cartography, as may clearly be seen in that masterpiece, the Catalan atlas of about 1375. Some attempts to extend the range of the portolan charts have already been noted, e.g. the chart of Angellino de Dalorto, and at about the same time Marino Sanudo was striving to reconcile the old and the new data. The completion of this reformation of the world map was the work of the Catalan cartographers.

Though the Catalan atlas is the earliest complete example of its kind which has survived, it was undoubtedly preceded by others of similar general design. The Medici sea atlas of 1351 contains a 'world' map (extending eastwards as far as the west coast of India only) which resembles it in the outline of the coasts and in interior details. From the nomenclature, it is probably of Ligurian origin. An even earlier chart (probably covering the whole 'world' originally), that by Angelino Dulcert, of Majorca, dated 1339, also has points of resemblance to the Catalan atlas of 1375. In view of the possible identity of Dulcert and Dalorto, and the Ligurian origin of the Medici atlas, we may conclude that this type of world map, though developed by Catalans, originated early in the fourteenth century in northern Italy, where the narrative of Marco Polo, which, as will be seen, supplied many of the details embodied in the map, would be most readily available.

We know in unusual detail the circumstances in which the Catalan atlas of 1375 (the date of the calendar which accompanies it) was produced and the career of the cartographer who compiled it. When in 1381 the envoy of the French king asked King Peter of Aragon for a copy of the latest world map (proof in itself that the reputation of the Catalan school had then been widely recognized) he was given this example, which has been preserved at Paris ever since. It is on record that it was the work of 'Cresques le juif'. Abraham Cresques, a Jew of Palma in the island of Majorca, for many years was "master of *mappae mundi* and of compasses", i.e. cartographer and instrument maker, to the King of Aragon, from whom he received special privileges and protection. There are several references to world maps executed by him, though this is the only one now known. After his death in 1387, his work was carried on by his son, Jafuda, but the day of the Jewish school of cartography at Majorca was already drawing to a close, owing to the wave of persecution which swept through the Aragon kingdom in the closing years of the century. Jafuda submitted to force and became a Christian in 1391, receiving the name of Jaime Ribes, but his position was not improved thereby, and he left Palma for Barcelona. Here he continued his work, in increasingly difficult circumstances, until finally he accepted the invitation of Prince Henry of Portugal to take up his residence in that country, where he instructed the Portuguese in cosmography and the making of charts. This link between the Majorcan school and the beginning of Portuguese overseas expansion is of obvious significance.

The patrons of Cresques, King Peter III of Aragon and his son, in addition to their scientific interests, were keenly interested in reports of Eastern lands, in relation to their forward economic policy, and were at special pains to secure manuscript copies of Marco Polo's 'Description of the world', the travels of Odoric of Pordenone, and, what may surprise the modern reader, the Voyage of Sir John Mandeville. Though fabulous in part, Mandeville's book has a scientific background. He was quite sound, for example, on the sphericity of the earth; as he says

". . . who so wold pursue them for to environ the earth who so had grace of God to hold the waye, he mighte come right to the same countreys that he were come of and come from and so go about the earth . . . fewe men assay to go so, and yet it might be done."[1]

The title of the atlas shows clearly the spirit in which it was executed and its content: "Mappamundi, that is to say, image of the world and of the regions which are on the earth and of the various kinds of peoples which inhabit it." The whole consists of twelve leaves mounted on boards to fold like a screen; four are occupied by cosmographical and navigational data, the remaining eight forming the map. Each leaf is 69 × 49 cms., so that the whole is approximately 69 cms. × 3.9 metres. These proportions are of some significance, for they have undoubtedly restricted the cartographer in his portrayal of the extreme northern and southern regions. This was

[1]Mandeville, Sir J., 'Voiage', Oxford, 1932, p. 247.

B*

perhaps to some extent deliberate—for two years before the composition of this map, we hear of the Infant John demanding a map "well executed and drawn with its East and West" and figuring "all that could be shown of the West and of the Strait (of Gibraltar) leading to the West". The Infant, in other words, was interested, not in northern Europe and Asia or in southern Africa, but in the Orient and the western Ocean. The cartographer satisfied him by cutting out, as it were, an east-west rectangle from a circular world map which would cover the desired area.

Later Catalan maps, e.g. the Este map, retained the circular form. The shape of the map, therefore, must not be taken as evidence on questions such as the extent or form of the African continent; nor does the change from a circular to a rectangular frame indicate specifically any change in ideas relating to the shape of the earth. As an astronomer, Cresques accepted its sphericity.

The sources of the Catalan Atlas fall into three groups: (1) elements derived from the typical circular world-map of medieval times; (2) the outlines of the Black Sea, Mediterranean, and the coasts of western Europe based on the 'normal' portolan chart; (3) details drawn from the narratives of the thirteenth- and fourteenth-century travellers in Asia, which transformed the cartographic representation of that continent.

The influence of the medieval world map may be seen in many features: Jerusalem, though not so strongly emphasized, is still approximately in the centre of the map; a portion of the original circumference of the circular map forms the coastline of north-east Asia, with the 'Caspian' mountains still enclosing the tribes of Gog and Magog; the large island of 'Taperobane' occupies approximately the same position as, for example, on the Hereford Map; the great west-east river beyond the Atlas Mountains resembles the traditional conception of the hydrography of North Africa, though contemporary names have been inserted. Clearly the contemporary additions are set in a much older framework.

The narratives of contemporary travellers were extensively used by the cartographer. The north-west coast of Africa extends beyond Cape Bojador to a point north of the Rio d'Oro. An inscription records the departure of the Catalan, Jacome Ferrer, on a voyage to this 'river of gold' in 1346, and some knowledge of the gold-producing region of the middle Niger is displayed. The regional name Guinea (Ginuia), the Kingdom of Melli, and stages on the routes from Morocco to the Niger, e.g. Sigilmessa, Tebelt, Tagaza and Tenbuch (Timbuktu), are marked.[1] In north-east Africa, a knowledge of the Nile valley as far south as Dongala, where there was a Catholic mission early in the century, is apparent. The delineation of the Nile system is vitiated, however, by the conception that it flowed from a great lake in the Guinea region. This lake may reflect rumours about the flood areas of the Niger, but the whole idea is very much older.

It is however in its representation of Asia that the greatest interest of the Catalan map lies. For the first time in medieval cartography, the continent assumes a recognizable form, with one or two notable exceptions. From the Caspian Sea in the west, with a fairly accurate outline in the style of the portolan charts, the Mongol domains stretch away eastwards to the coast of Cathay. This makes a sweep from east to south with an approach to its actual form, and along it appear several of the great medieval ports and trading centres, frequented by Arab merchants. In the interior are correctly placed the main divisions of the Mongol territory; from west to east, the 'Empire of Sarra' (the Kipchak khanate), 'the Empire of Medeia' (the Chagtai khanate of the middle), and the suzerain empire of the Great Khan, Catayo with its capital at Cambaluc (Peking).

If the map is stripped of its legends and drawings of the older tradition, it is apparent that the main interest of the compiler is concentrated in a central strip across the continent. Herein lies a succession of physical features—mountain, river, and lake—and of towns with corrupt but recognizable forms of their medieval names as given in the narratives of the great travellers of the thirteenth century. These are jumbled together in a manner sometimes difficult to understand, but with the help of Marco Polo's narrative, it is possible to disentangle the

[1]On this traffic see Taylor, E. G. R., 'Pactolus, river of gold' (*Scottish Geogr. Mag.*, 44, 1928, 129) and 'The Voyages of Cadamosto', *Hakluyt Soc.*, 2nd series, vol. 80, 1937.

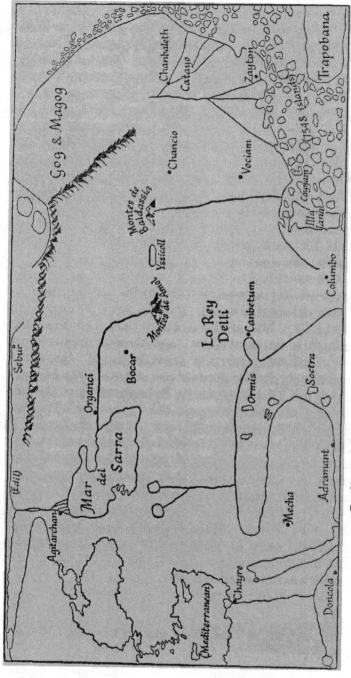

Outline of the eastern section of the Catalan Atlas, c. 1375

itineraries which the map was evidently intended to set out.

In the west is the Oxus river (fl. Organci) shown, as on most contemporary maps, flowing into the Caspian, and alongside it the early stages of the itinerary, from Urganj (the medieval Khiva) through Bokhara and Samarcand to the sources of the river in the mountains of Amol, on the eastern limits of Persia. These are the highlands of Badakshan where the route crossed the Pamirs. East of this lies the lake Yssikol and Emalech the seat of the Khan, the Armalec of other travellers, in the Kuldja region. The delineation is then confused by the repetition of the Badakshan uplands, the mountains of 'Baldassia', a mistake which probably arose from a confusion over the river system of southern Asia. Further east is 'Chancio' (Kanchow, on the great loop of the Hwang-ho), and ultimately 'Chambaleth', the city of the Great Khan, and the goal of travellers from the west.

This, with several omissions, was in outline the route followed by Marco Polo's father and uncle on their first journey to the Great Khan's court. It is also possible to discern traces of their second journey, accompanied by Marco, on a more southerly route through Eri (Herat), Badakshan, and along the southern edge of the Tarim basin from Khotan to the city of Lop. The compiler, however, perhaps because he confused this desert area with the Gobi, has transferred this stretch to the north of the Issik Kul.

A third route is indicated rather confusedly on the extreme northern edge of the map. It is marked by a line of towns up the valley of the Volga from 'Agitarchan' (Astrakhan) through 'Sarra' (Sarai), 'Borgar', and thence eastwards through 'Pascherit' (probably representing the territory of the Bash Kirds east of the middle Volga), and 'Sebur', or Sibir, a medieval settlement whose site is unknown, but thought to be on the upper Irtish. In this quarter, the information on which the map was based was not drawn from Marco Polo. To the south is a long east-west range, called the 'mountains of Sebur', representing the north-western face of the Tien Shan and Altai. In the late thirteenth and early fourteenth centuries there were Franciscan mission stations at these localities, and the details no doubt came originally from the friars.

'Chambaleth', the city of the Great Khan which so intrigued the chroniclers of the fourteenth century, receives due prominence, with a long legend describing its magnitude and grandeurs. It stands near the apex of a triangle formed by two rivers and the ocean; each of the two rivers divides into three before reaching the sea, a representation embodying a somewhat confused notion of the interlinked natural and artificial waterways of China.

On the southern portion of the Cathay coast, the general uniformity of the coast is broken by three bays, and it is significant that these are associated with the three great ports, Zayton (near Changchow), Cansay (better known in medieval records as Quinsay, i.e. Hangchow) and Cincolam (Canton). Of these, Canton is not mentioned by Marco Polo; it was however much frequented by Arab navigators and traders, upon whose reports the compiler was probably drawing. The attempt at representing the configuration of the coast suggests at least that his informants were interested from a maritime point of view. Some of the islands off Quinsay may stand for the Chusan archipelago, and further to the south is the large island of Caynam, i.e. Hainan. In the interior, the towns, in Cordier's view, can in general be related to the itineraries described by Polo.

South-east of the coast of Cathay are numerous islands— we are told that they number 7,548—in which grow the spices. In the extreme corner is a portion of a great island which is named Taprobana. A legend states that it is the last island in the East, and is called by the Tatars 'Great Caulij'. Yule pointed out that Kao li was the name for Korea, and he therefore considered that the island depicted confused notions of the Korean peninsula and Japan.

The delineation of the coastline of southern Asia has one major defect and one outstanding merit; the defect is the entire omission of the south-eastern peninsula; the merit is the portrayal for the first time of the Indian sub-continent in its peninsular form. The first is difficult to explain; to make up for it the cartographer has inserted a great island of Java (mis-spelt Jana), which however was probably intended for Sumatra. For the Indian peninsula, other sources are intermingled with Polo's account. The Kingdoms of India as enumerated by Polo are absent from the map, and there are significant differences in the towns appearing in the two documents. Conspicuous on the map is the 'Christian kingdom' and city of 'Columbo', placed on the east coast. There is no doubt however that this is Quilon, on the west coast. This form of the name (it is rendered Coilum by Polo), and other details, suggest that the compiler drew upon the writings of Friar Jordanus, who was a missionary in this area, and whose 'Book of Marvels' was completed and in circulation by 1340. In the area around the Gulf of Cambay, several towns are shown which are mentioned by Jordanus but not by Polo, e.g. Baroche and Gogo. There are other names, however, which are not found in Jordanus; but the commercial importance of Cambay (Canbetum, on the map), would account for the relatively detailed information about this region. There is, however, no indication of the great river Indus, a striking omission also from Polo's narrative. This probably arose from confusion between the Indus and the Ganges.

For the portion of the Indian Ocean included in the map, sources other than those embodied in Polo have been used. The Persian Gulf, extending almost due west, has an outline similar to that on the Dulcert map, but is otherwise superior to any earlier map. In the Gulf, the 'island of Ormis' (Hormuz) is shown, opposite the former settlement of the same name on the mainland. The Southern Arabian coast has names differing from those given by Polo, and in one of them 'Adramant' we may recognize the modern Hadhramaut. The island of 'Scotra', an important stage on the trade route from Aden to India, is misplaced to the east, and appears to occupy the approximate position of the Kuria Muria islands.

For India and the ocean to the west, therefore, we may conclude that charts were used which differed in detail from Polo's account, though similar in general features. That such charts existed we know from Polo's own statements. Possibly additions were also made so that the map might serve as an illustration to his narrative.

The only complete Catalan world map other than that of 1375 which has survived is the Este map preserved at Modena.

This map is circular, and although almost a hundred years later, it is clearly related to the Atlas of 1375. This resemblance in the content of the two maps strengthens the contention that the latter was derived from a circular prototype. The nomenclature and the numerous legends, mostly in Catalan with a few in corrupt Latin, are often very similar to those of the 1375 Atlas. In some instances the legends are more complete, in others they are less detailed; they suggest therefore not direct copying but a common source. This similarity is also evident in the delineation of the main features—most of those in the 1375 Atlas are to be found on the Este map.

The northern portions of Asia and Europe, which lay outside the limits of the Catalan Atlas, significantly, contain very little detail. On the southern coastline of Asia there are some differences, generally slight, between the two maps. The peninsula of India is much less pronounced on the Este map, and to the south is the large island of 'Salam' or 'Silan' (Ceylon) which fell outside the limits of the Catalan Atlas. A legend refers to its wealth in rubies and other precious stones. There can be no doubt however that the two outlines are fundamentally identical. To the east is the island of 'Java', as on the Catalan Atlas. The island of 'Trapobana' is much enlarged, and is placed on the south-eastern margin of the map. The surrounding ocean, the 'Mar deles indies' is filled with numerous nameless and featureless islands.

Africa occupies most of the southern half of the map. The continent ends in a great arc, conforming to the circular frame of the map, and extending eastwards to form the southern boundary of the Indian Ocean. On the west, a long narrow gulf from the circumfluent ocean almost severs this southerly projection from northern Africa. The southern interior is blank save for the legend "Africa begins at the river Nile in Egypt and ends at Gutzola in the west: it includes the whole land of Barbaria, and the land in the south". This outline and legend have been interpreted to imply some knowledge of the southern extremity of Africa, and perhaps of a practicable route from the west to the Indian Ocean.

That the great western gulf reflects some knowledge of the Gulf of Guinea is more probable. The design of the northern half of the continent in general resembles that of the other Catalan charts, but the north-western coast embodies some details of contemporary Portuguese voyages as far as 'C. ŭde' (Cape Verde) and 'C. groso'. From this evidence, the map is usually dated about 1450. Near the gulf are the Mountains of the Moon, from which five rivers flow northwards to a lake on the 'western Nile'. This lake probably represents the area around the Upper Niger liable to inundation; Dr. Kimble has pointed out that these rivers may well represent the five main sources of the Niger. These Mountains of the Moon are stated to be on the Equator, and the streams are called the 'riu de lor'. We may therefore assume that the headwaters of the Niger marked the approximate limit of knowledge in this region, and it is not improbable that reports of the sea to the south had been received. These may have induced the cartographer to accept the western gulf of Ptolemy, but to enlarge it considerably. The name 'river of gold' recalls the inscription on the Catalan Atlas. The portrayal of the interior thus goes back at least to 1375. Apart therefore from a small portion of the coastline, the map owes nothing to Portuguese exploration.

Some surprise has been expressed that a map of 1450 should contain relatively up-to-date details with antiquated ideas in other areas, and this has produced some rather involved explanations. Taking into consideration the lack of details and names in the southern regions of Africa, we may plausibly conjecture that, as an exception to the usual conservatism, the draughtsman, in Africa at least, had removed all the detail for which he had no evidence, to obtain a framework on which to insert the latest Portuguese discoveries. It must remain debatable whether the outline of the southern extremity represents some knowledge of the Cape. The outline may be entirely imposed by the frame of the map: at the most, it may reflect the kind of report that we find on Fra Mauro's map.

The merit of the Catalan cartographers lay in the skill with which they employed the best contemporary sources to modify the traditional world picture, never proceeding further than the evidence warranted. In the same spirit they removed from the map most of the traditional fables which had been accepted for centuries, and preferred, for example, to omit the northern and southern regions entirely, or to leave southern Africa a blank rather than to fill it with the *anthropagi* and other monsters which adorn the medieval maps. Though drawings of men and animals still figure on their works they are in the main those for which there was some contemporary, or nearly contemporary, warrant; for example, Mansa Musa, the lord of Guinea, whose pilgrimage to Mecca created a sensation in 1324, or Olub bein, the ruler of the Tatars. In this spirit of critical realism, the Catalan cartographers of the fourteenth century threw off the bonds of tradition, and anticipated the achievements of the Renaissance.

REFERENCES

BEAZLEY, C. R., The dawn of modern geography. Vol. 3. 1906

BUCHON, J. A. C., and J. TASTU, Notice d'un atlas en langue catalane, 1375. *Paris*, 1839.

CORDIER, H., L'Extrême-Orient dans l'Atlas Catalan de Charles V. (*Bull. de géogr. hist. et descr. Paris*, 1895.)

KIMBLE, G. H., The Catalan world map of the R. Biblioteca Estense at Modena. (With fascimile.) R.G.S., London, 1934.

KRETSCHMER, K., Die Weltkarte der Bibliotheca Estense in Modena. (*Zeits. Ges. f. Erdkunde, Berlin*, 1897.)

REPARAZ, G. DE, L'activité maritime et commerciale du royaume d'Aragon au XIIIe siècle et son influence sur le développement de l'école cartographique du Majorque. (*Bull. Hispanique*, 49 (1947) 422–51.)

———— Les sciences géographiques et astronomiques du XIV* siècle dans le nord-est de la Péninsule Ibérique. (*Archives internat. d'histoire des sciences* 3 (1948).)

YULE, Sir H., and H. CORDIER, The book of Ser Marco Polo, 3rd ed. 1903.

———— Cathay and the way thither. New ed. 4 vols. (*Hakluyt Soc. ser.* ii, v.33, 37, 38, 41.) 1913–16.

CHAPTER IV

FIFTEENTH-CENTURY WORLD MAPS: FRA MAURO AND MARTIN BEHAIM

CONTEMPORARY with the later Catalan maps are several mainly of Italian origin which also preserve some medieval features, but show very markedly the influence of Ptolemy's 'Geography', manuscripts of which were circulating in western Europe at least as early as the opening decades of the fifteenth century.

An early, but not very successful, attempt to reconcile the classical and medieval outlook is the world map drawn by the Benedictine Andreas Walsperger at Constance in 1448. "In this figure," he writes, "is contained a mappa mundi or geometrical description of the world, made out of the Cosmographia of Ptolemy proportionately to the latitudes, longitudes and the divisions by climates, and with the true and complete chart for the navigation of the seas." He has not made, to say the least, the best of his authorities, and the resulting map is muddled and difficult to explain. There are one or two interesting points; e.g. the use of red dots for Christian and black dots for infidel cities; also the orientation, with the south at the head. Though the east includes the terrestrial paradise, represented by a great Gothic castle, there are some glimmerings of recent knowledge. The Indian sea is not closed, but connected by a channel with the ocean. The island 'Taperbana' is inscribed 'the place of pepper', and an unnamed island off the Arabian coast (perhaps Ormuz or Socotra) has the legend 'Here pepper is sold'. Such details point to an interest in the spice trade before the Conti-Bracciolini report. The contrast between this map and the elliptical world map of 1457, preserved in the National Library, Florence, is striking. The latter, usually considered to be of Genoese origin, is very carefully drawn, particularly the outline of the Mediterranean. It has a number of neatly executed drawings, and legends in

Latin. Unlike most maps of this type, it has a scale, each division of which represents 100 miles. The title is rather difficult to decipher and recalls Walsperger's. An approximate translation is: "This is the true description of the world of the cosmographers, accommodated to the marine (chart), from which frivolous tales have been removed."

The elliptical frame is unusual for this period, but it appears to have no great significance. The outline, particularly in Asia, is largely Ptolemaic. After the Alexandrian, the second main authority for the eastern portion is Nicolo Conti, the Venetian traveller, who reached the east Indian islands and perhaps southern China, and whose narrative was written down by Poggio Bracciolini shortly after 1447.

The details from Conti's narrative make a considerable showing: e.g. the large lake in India between Indus and Ganges "of a marvellous sauerie and pleasaunt water to drink, and al those that dwell there about drink of it, and also farre off . . .";[1] the island 'Xilana' (Ceylon) to the east of the peninsula; the great city 'Biznigaria', representing the Vijayanagar kingdom of southern India, which occurs in most late fifteenth-century accounts, but here sadly misplaced near the Ganges; the details of the nature of the Ganges delta; the addition of 'Scyamutha' (Sumatra) as an alternative name for Taprobana. The name Sine, for China, was also probably taken from Conti.

But it is perhaps in respect to the islands of the south-east that the map is of greatest interest. In the extreme east are two large islands, Java major and Java minor, and to the south-east two smaller islands 'Sanday et Bandam'. All these are taken from the Conti narrative: Java major is thought to be Borneo, and Java minor the island now known by that name. Though the names Sanday and Bandam have not been satisfactorily explained, the reference in the legend to spices and cloves makes it fairly certain that they are islands of the Molucca group. If this is so, this is the first time that the much sought after spice islands appear clearly on a map. Conti describes them as lying on the extreme edge of the known world: beyond

[1]Quoted from Conti's Elizabethan translator, Frampton.

them navigation was difficult or impossible owing to contrary winds. In the southern sea there is a note: "In this sea, they navigate by the southern pole (star), the northern having disappeared." This also is taken straight from Conti.

The main African interest lies in the fact that, as a departure from Ptolemy's conception, the Indian Ocean is not land-locked, and, significantly, the southern extremity of Africa does not run away eastwards, as on the Este map. At first sight, it is not clear that Africa is completely surrounded by the ocean, but closer examination shows that the blue of the ocean and the red on the land have faded, and that a definite coastline had been originally drawn in.

This map has recently attracted attention by the claim of S. Crinò that the famous chart which Toscanelli sent to the King of Portugal in 1474, and later but less certainly to Columbus, was a copy of it. Crinò claimed that it is of Floren-tine, not Genoese, origin; that the style of writing and certain other features definitely indicate that it was drawn by Tos-canelli; and that it agrees closely with the letter sent to Portugal with the copy, so closely in fact that the letter is merely a commentary upon it. All these arguments, and many more, have been warmly, even acrimoniously, contested. Without an expert and minute palaeographical investigation, it is impossible either to accept or reject the attribution to Toscanelli, but Crinò presented a case which requires further examination. On the question, of main interest here, as to the correspondence between the letter of 1474 and the map of 1457, it is possible, however, to form some opinion. The main objection to Crinò's thesis is that the letter definitely refers to a chart for navigation, while the 1457 map is primarily a world map drawn by a cosmo-grapher. Further, the Toscanelli chart presumably depicted the ocean intervening between the west coast of Europe and the 'beginning of the East'. On the map of 1457, this ocean is split into two, and falls on the eastern and western margins. Though Crinò raised many points of interest, he did not establish his case beyond reasonable doubt. Biasutti argued that the horizon-tal and vertical lines on the map are parallels and meridians taken from the world map of Ptolemy, and that the longitudinal extent of the old world approximately corresponds to his figure of 180°. It is difficult to see, therefore, if this map of 1457 was similar to that sent to Portugal, where its importance lay, for this information was accessible to all inquirers. The interest of the cartographer seems more probably to have lain in Conti's description of the oriental spice islands and the possibility of reaching them by circumnavigating Africa. His work is clearly related, though not closely, to the great map of Far Mauro, his contemporary.[1]

The world map of Fra Mauro, a monk of Murano, near Venice, is often regarded as the culmination of medieval cartography, but in some respects it is transitional between medieval and renaissance cartography. Fra Mauro appears to have had a considerable reputation as a cartographer, and to have been at work on a world map as early as 1447. Ten years later, he was commissioned by the King of Portugal to construct another, and for this purpose he was provided with charts showing the latest discoveries of the Portuguese (according to an inscription off the west coast of Africa). In this work he was assisted by the chart maker, Andrea Bianco, the draughtsman of a world map dated 1436, and a number of illuminators. The map for the King, finished in April 1459, was sent to Portugal, but cannot now be traced. Fra Mauro died shortly after, while working on a copy destined for the Seignory of Venice and completed later in 1459. This copy has survived and is now preserved in the Marciana library at Venice. The map is circular, its diameter approximately 6 feet 4 inches, and is drawn on parchment mounted on wood. It is full of detail, carefully drawn and coloured, and annotated with very many legends. Though the coasts are drawn in a style recalling that of the portolan charts, loxodromes and compass roses are absent, and the effect is definitely that of a *mappa mundi*, not a nautical chart, especially as it is oriented with the south at the head.

The convention of placing the centre of the map at Jerusalem has at last been abandoned: perhaps under the direct influence of Ptolemy, or of travellers' reports on the great extent of the

[1]For statements of the conflicting views of S. Crino, R. Biasutti and A. Magnaghi, see *Revista geographica italiana*, 49, 1942, 35–54, where there are also numerous references to the literature on the controversy. H. Winter also pronounced against Crino's thesis ('Die angebliche Toscanelli-Karte' *Koloniale Rundschau*, 33, 1942, 228–38.)

East. This departure from orthodox practice clearly worried the friar, and he excuses himself thus:

"Jerusalem is indeed the centre of the inhabited world latitudinally, though longitudinally it is somewhat to the west, but since the western portion is more thickly populated by reason of Europe, therefore Jerusalem is also the centre longitudinally if we regard not empty space but the density of population."

It is clear from numerous legends that Fra Mauro was very much aware of the great deference then paid to the cosmographical conceptions of Ptolemy, and the likelihood of severe criticism for any map which ignored them. Nevertheless, in general, he stands by contemporary ideas and forestalls criticism thus:

"I do not think it derogatory to Ptolemy if I do not follow his *Cosmografia*, because, to have observed his meridians or parallels or degrees, it would be necessary in respect to the setting out of the known parts of this circumference, to leave out many provinces not mentioned by Ptolemy. But principally in latitude, that is from south to north, he has much 'terra incognita', because in his time it was unknown."

If Fra Mauro's basis was less scientific than it might have been, he did at least point to the necessity for modifying Ptolemy's ideas in the light of more recent knowledge. In one major modification, the opening of the 'Sea of India' to the circumfluent ocean, he was in accord with all his contemporaries. Ptolemy, he writes, like all cosmographers, could not personally verify everything that he entered on his map and with the lapse of time more accurate reports will become available. He claimed for himself to have done his best to establish the truth.

"In my time I have striven to verify the writings by experience, through many years' investigation, and intercourse with persons worthy of credence, who have seen with their own eyes what is faithfully set out above."

He also displays a critical spirit when he inserts in the far northeast of Asia, near the enclosed tribes—"I do not think it possible for Alexander to have reached so far"—and expresses his doubts about these mountains really being the Caspian range; or when he writes "Note that the Columns of Hercules mean naught else than the break in the mountains which enclose the Strait of Gibilterra".

He had not been able to arrive at an opinion on the size of the globe:

"Likewise I have found various opinions regarding this circumference, but it is not possible to verify them. It is said to be 22500 or 24000 *miglia* or more or less according to various considerations and opinions, but they are not of much authenticity, since they have not been tested."

He had therefore no very accurate knowledge of what proportion of the earth he was portraying in his map. By moving its centre eastwards, however, he had made the relative longitudinal extents of Europe and Asia approximately correct. Putting the centre at Jerusalem had of course resulted in the longitudinal extent of Asia being reduced in relation to that of the Mediterranean: on his map he represents it as about twice the length of that sea, which is fairly accurate for that latitude.

Having enlarged Asia in relation to Europe, our cosmographer has not put the additional space to very good use. It is extremely difficult to comprehend his representation of southern Asia. From the Persian Gulf eastwards, he appears to have taken the Ptolemaic outline, but exaggerated the principal gulfs and capes, and to this outline he has fitted the contemporary nomenclature. The great Gulf of Cambay recalls the similar feature of the fourteenth-century charts, with the addition of the island of Diu, an important trading centre. It is noticeable here that the order of the names from Gogo to Tana is reversed, probably an error in compilation due to the unusual orientation of the map. Beyond Tana, India is broken into two very stumpy peninsulas, resulting in the confusion of relative positions in the interior, and in the placing of Cape Deli in the latitude of Cape Comorin. Seilan (Ceylon) appears more or less correctly related to Cape Comorin, with a note that Ptolemy had confused this island with Taprobane, and a representation of Adam's Peak. To the east, there is a more or less recognizable Bay of Bengal, confined on the further side by the great island of Sumatra. Into this bay flows in the north a great river here named Indus, the repetition of a mistake which seems to go back at least as far as the Catalan Atlas. There is nothing corresponding to the Golden Chersonesus or the Malay Peninsula, but to the east again, rather surprisingly, is placed the 'Sinus Gangeticus', with the Ganges entering on the north: that river is therefore brought into close relationship with southern China. A conspicuous feature in the Indian Ocean is the Maldive Islands, shown with their characteristic linear extension. Instead however of running north and south, they extend approximately from north-west to south-east, and this direction is emphasized in an inscription. The position in which the Andaman Islands are shown in relation to Sumatra also suggests that there is a general tilting of the map in this area of about 45° west of north. In the south-east close to the border of the map is an island with the inscription "Isola Colombo, which has abundance of gold and much merchandise, and produces pepper in quantity. . . . The people of this island are of diverse faiths, Jews, Mahomedans and idolaters. . . ." This refers to the district of Quilon (the 'Colombo' of the Catalan Atlas) in the south of the Indian peninsula. Arab topographers were accustomed on occasion to refer rather loosely to districts approached by sea as 'islands' (*gezira*) and this often led to confusion, as in the present instance. This error suggests that portions of the map were probably based on written descriptions or sailing instructions by Arab merchants or pilots. Fra Mauro, or the draughtsman of his prototype, clearly misunderstood the passage referring to 'Colombo'. The notes attached to some of the islands, giving their direction in relation to others, as in the case of the Maldives already quoted, support this probability. Certainly the whole southern outline of the continent as depicted here could scarcely have been taken directly from a chart drawn by a practical navigator.

To the east of the Bay of Bengal is a very large Sumatra, the first time that name appears unequivocally on a map. To the north of it, and somewhat squeezed together by the limit of the map are many islands. As Fra Mauro states that in this region lack of space had compelled him to omit many islands, it no doubt also obliged him to alter their orientation drastically. A long legend here gives some illuminating details on the traffic in spices and pepper.

"Java minor, a very fertile island, in which there are eight kingdoms, is surrounded by eight islands in which grows the 'sotil specie'. And in the said Java grow ginger and other fine spices in great quantities, and all the crop from this and the other (islands) is carried to Java major, where it is divided into three parts, one for Zaiton (Changchow) and Cathay, the other by the sea of India for Ormuz, Jidda, and Mecca, and the third northwards by the Sea of Cathay. In this island according to the testimony of those who sail this sea, the Antarctic Pole star is seen elevated at the height of *un brazo*." (This term has never been satisfactorily explained.)

Java major is said to be especially associated with Cathay:

"Java major, a very noble island, placed in the east in the furthest part of the world in the direction of Cin, belonging to Cathay, and of the gulf or port of Zaiton, is 3,000 miles in circumference and has 1,111 kingdoms; the people are idolatrous, sorcerers, and evil. But the island is all delightful and very fertile, producing many things such as gold in great quantities, aloes wood, spices, and other marvels. And from the Cavo del ver southwards there is a port called Randan, fine, large, and safe: in the vicinity is the very noble city Java, of which many wonders are told."

The islands south of Java minor doubtless represent the Moluccas, as on the Genoese map. There is one tantalizing point: just to the north of Java major is a small island 'isola de Zimpagu'. Can this be Cipangu (Japan), and thus the first appearance of the name on a map? It is certainly far from its correct position, but, as the cartographer has had to omit many islands for lack of room and doubtless pressed others together, this name may easily have been misplaced. If 'Java major' is not Java, but another island closer to Zaiton, the possibility is greater. All this information on the spice islands and their trade is taken from the Conti document.

For the representation of China, a great deal has been drawn from Marco Polo's narrative, as for the Catalan Atlas. Fra Mauro's delineation however differs from that of the latter in two respects: the coast of China is broken by several long and narrow gulfs, which on inspection are seen to be merely over-emphasized estuaries or important ports such as Zaiton. Of more interest is the improved hydrographic system. Instead of the rivers radiating from a point near Cambalec, the two principal rivers are shown with some approach to reality. The upper course of the Quiam (the Yangtse Kiang), "the greatest river in the world", it is true, is brought too far south, but the Hwang ho has its great upper bend clearly drawn. (There is no question, of course, of these rivers being drawn 'true to scale'.)

The towns, and the numerous annotations, are taken directly, it would appear, from Polo's narrative. Most of those, for instance, which occur in his itinerary from Cambalec to Zaiton, are to be found on the map, though in no very comprehensible order, often accompanied by a drawing of a feature mentioned by Polo, or his comments, e.g. on the gold and silk of this city, or the porcelain of that; the sugar for which this district is noted or the gigantic reeds which grow in another. In the western regions, the picture is confused owing to the inadequate space allotted to them. Fra Mauro seems to have been interested in Persia and Mesopotamia and to have drawn maps of these countries before beginning his world map. This probably explains why they figure so conspicuously on the latter, at the expense of the features of eastern Asia. Thus the Issik Kul, approximately in its correct relative position on the Catalan Atlas, is shown almost neighbouring on Cambalec, and other places, Armalec and Hamil, for instance, have been similarly displaced. As on the Catalan Atlas, the kingdom of Tenduc has been relegated to the north, in proximity to the 'enclosed tribes'.

On the whole, however, a fair knowledge of China is displayed; the mid-nineteenth century certainly knew less of the interior of Central Africa than the fifteenth century did of the interior of China.

Yule believed that Conti had probably supplied Fra Mauro verbally with information on south-east Asia, additional to that contained in his published narrative. In Burma, for instance, there are the cities of Perhé (the correct Burmese form), Pochang (Pagán, the ancient capital) and Moquan (Mogoung). In the upper course of the Irrawaddy there is a note testifying to knowledge of commercial routes: "Here goods are transferred from river to river, and so go on into Cathay."

India is also rich in towns, but for the reasons already discussed, their relative positions are faulty. Orica, Sonargauam, and Satgauam (Satganev), all in the Ganges delta, are probably due to Conti. Goa, later to become the centre of Portuguese power in India, is entered under its earlier name of Boa Zandapur.

Africa in outline resembles the representation on the Este map, save that it is not almost severed in two by the prolongation of the 'Sinus Ethiopicus'. Details of Abyssinian topography have been expanded to cover most of the centre and south, except for the southernmost extremity, which is separated by a river or channel from the main, and named 'Diab'. The detailed knowledge of the north-east African interior extends as far as the river Zebe (?Webi Shebeli). The Nile (Blue Nile) is shown rising near a lake, undoubtedly Lake Tana, in the fountain of Geneth, a name for the source which was still in use in James Bruce's time, more than three hundred years later. Fra Mauro states that he obtained this information from natives of the country "who with their own hands had drawn for me all these provinces and cities, rivers and mountains, with their names—all of which I have not been able to set down in proper order from lack of space". It has been shown that two main causes of the confused representation of north-east Africa are the ignorance of the cartographer about the existence of the eastern Sudan, so that he telescoped Egypt and Abyssinia together, and the failure to realize that much of the hydrographic detail available applied to one river only, the Abbai, and not to a number of distinct streams.

The Coptic Church of Abyssinia was in touch with Cairo and Jerusalem, and it was doubtless from emissaries of the Church that Fra Mauro obtained his information. Near Lake Tana he has the name 'Ciebel gamar', literally 'mountain of the moon'. Mr. O. G. S. Crawford suggests that this was the origin of the legend about the source of the Nile, and that it was only later that the site was transferred to the Equator.[1]

The suggestion is partly retained of a 'western Nile' flowing from a great marsh, no doubt Lake Chad; beyond this marsh a river flows westwards to enter the ocean by two branches to the north of Cape Verde, no doubt the Senegal and perhaps the Gambia. Fra Mauro tells us that he was supplied with Portuguese charts and had spoken with those who had navigated in these waters. Actually the only contemporary names he has are 'C. Virde' and C. Rosso, immediately north of the great gulf; the small river in the vicinity may be the Rio Grande. The drawing of the coastline does not show much correspondence with reality. The Portuguese are stated to have reached the meridian of Tunis and perhaps even that of Alexandria. Curiously enough, on the map the eastern end of the gulf may be said to be on the meridian of Tunis, as in fact the eastern terminus of the Gulf of Guinea is. (To have crossed the meridian of Alexandria, however, would have entailed rounding the Cape of Good Hope.) By 1459 the Portuguese navigators had probably not passed beyond Sierra Leone, and it is disputed whether at that date the Cape Verde Islands had been discovered. The delineation of the great gulf can scarcely rest on first-hand knowledge possessed by the Portuguese. The lack of the latest information on the map has been criticized, especially as Bianco was employed in its production, but it is scarcely justifiable to argue from this that

[1]Crawford, O. G. S., 'Some medieval theories about the Nile' (*Geogr. Journ.*, 114, 1949, 6–29.)

information was deliberately withheld from the cartographer by the Portuguese authorities. They, after all, were well informed on the progress of their navigators. In causing the world map to be drawn, they were presumably interested in the sea-route round Africa to the Indies, and as we have seen, the latest information on the spice islands was incorporated in it.

On the southern island, 'Diab', already mentioned, there are a number of names, including 'Xēgibā' (Zanzibar), 'Soffala' 'Chelue' (Kilwa) and 'Maabase' (Mombasa). These names are of Arab origin, and Arabs had been active on this coast for centuries. The strength of tradition and its influence on European cartographers is strikingly illustrated in a legend placed near the southern extremity which has attracted much attention. It reads

"About the year of Our Lord 1420 a ship or junk of India on a crossing of the Sea of India towards the islands of men and women was driven beyond the Cape of Diab and through the Green Islands and the darkness towards the west and south-west for forty days, finding nothing but air and water, and by their reckoning they ran 2,000 miles and fortune deserted them. They made the return to the said Cavo de Diab in seventy days and drawing near to the shore to supply their wants the sailors saw the egg of a bird called *roc*, the egg being as big as a seven gallon cask, and the size of the bird is such that from the point of one wing to another was sixty paces and it can quite easily lift an elephant or any other large animal. It does great damage to the inhabitants and is very fast in its flight."

(Elsewhere he says he had spoken to persons who had been driven forty days beyond the Cavo de Soffala.) The *roc* is of course the fabulous bird of the 'Arabian Nights'. But the interesting point is that, five hundred years before Fra Mauro's time an Arab chronicler writing about Sofala has a very similar story of a vessel not only being driven by storm but also encountering the *roc*. Fra Mauro was here drawing ultimately on Arabic sources, and the doubt arises whether any significance should be attached to the date of 1420. There is other evidence of eastern sources in this quarter: for instance, the names of the two islands Negila (*Sanskrit*, beautiful) and Mangula (*Arabic*, fortunate).

The island of Diab is probably based on reports of the existence of the great island of Madagascar. There would be no improbability in a vessel being driven down to the latitude of the Cape of Good Hope, or of Arabs at Soffala having some inkling of the trend of the coast to the south. It is extremely unlikely, as has been argued, that the Cape of Diab is nothing more southerly than Cape Guadafui. Fra Mauro himself certainly accepted the possibility of circumnavigating southern Africa.

On this and other evidence, Fra Mauro reached an important conclusion:

"Some authors state of the Sea of India that it is enclosed like a lake, and that the ocean sea does not enter it. But Solinus holds that it is the ocean, and that its southern and south-western parts are navigable. And I affirm that some ships have sailed and returned by this route."

This map is of special interest as showing that, at least forty years before the Portuguese reached India, Arab sailing directions covering the east coast of Africa, India, and the seas beyond to the vicinity of Sumatra, or at least information derived from such sources, were available in western Europe. Taken as a whole, the map can have offered nothing but encouragement to the Portuguese to persevere.

By the time Fra Mauro was working on his map, the known world was expanding beyond the conventional framework of the circular *mappa mundi*. This expansion was both to the east and to the west, to Cathay in the east, and to the Atlantic islands in the west. If the diameter of the map was increased to accommodate these new details, the northern and southern quadrants, correspondingly enlarged, looked more empty than ever. In Bianco's world map of 1436, the continental mass is placed excentrically to the embracing ocean, and eastern Asia breaks through the framework in order to leave more space in the west for the insertion of Antillia.

As we have seen, Cresques had abandoned the circular form a century earlier. When it became apparent that Jerusalem could no longer be regarded literally as the centre of the known world, the arguments for a circular frame lost much of their force. Further, the popularity of Ptolemy's world map also worked in this direction, apart from the fact that, without considerable knowledge of mathematics, it was impossible to fit meridians and parallels satisfactorily into a circle, that is, to construct a precise projection. With this world map of Fra Mauro, therefore, we leave the medieval convention which had prevailed for so many centuries. The last important pre-Columbian representation of the world was in fact a globe, the earliest to have survived.

Martin Behaim's globe

The main features of interest in the Behaim globe are first the fact that it is a globe and that the maker was therefore obliged to consider directly the width of the ocean between Europe and Asia; second, the strong probability that the outlines adopted on the globe, with the exception of the African coast, were taken from a printed map already fairly widely circulated; third, the persistency with which these outlines were adhered to by later cartographers and their determined efforts to force the new discoveries into this framework. The globe has also great importance in the perennial controversy over the initiation of Columbus' great design and the subsequent evolution of his ideas on the nature of his discoveries, though a detailed discussion of these problems lies outside the present study.

The former fame of Martin Behaim as a skilled cosmographer has now faded. Ravenstein has shown that Behaim possibly made a voyage to Guinea in 1484-5, but that he was certainly not an explorer of the southern seas and a possible rival of Columbus, and his cartographical attainments were distinctly limited. All the available evidence tends to show that he was a successful man of business who made a certain position for himself in Portugal, and who, like many others of his time, was keenly interested in the new discoveries.

In the year 1490 Martin Behaim returned to his native city

of Nuremburg for a stay of three years and it was then, at the
request of influential burghers that the globe was made.
Behaim received payment for "a printed *mappa mundi* embracing the whole world", which was used in making the globe.
Since he is said to have "expended thereon his art and pains",
he may be credited at least with amending the printed representation in the African section, though his contribution was
not distinguished. As far as one can tell from a facsimile, the
drawing and illumination of the globe's surface were carefully
and attractively executed; for this the credit must go to the
miniaturist, Georg Holzschuher.

The globe is twenty inches in diameter: on it appear the
Equator, the two tropics, and the Arctic and Antarctic circles.
The Equator is divided into 360 degrees, but these are unnumbered. One meridian, 80° to the west of Lisbon is shown,
and this is likewise graduated for degrees. These are also
unnumbered, but in high latitudes the lengths of the longest
days are given. The longitudinal extent of the old world
accepted by Ptolemy was approximately 177° to the eastern
shore of the Magnus Sinus, plus an unspecified number of
degrees for the remaining extent of China. Behaim accepted
more or less Ptolemy's 177° and added 57° to embrace the
eastern shores of China. He thus arrived at a total of 234°, the
correct figure being 131°. The effect of this was to reduce the
distance from western Europe westwards to the Asiatic shores
to 126°, in place of the correct figure of 229°. There is no
indication on the globe of what Behaim considered the length
of a degree to be—but even if he did not go as far as Columbus
in adopting the figure of 56⅔ miles for a degree, he presented
a very misleading impression of the distance to be covered
in reaching the east from the west. Since in addition, Cipangu,
in accordance with Marco Polo's report, is placed some 25°
off the coast of China on the tropic of Cancer, and the Cape
Verde Islands are shown as extending to 30° west of the Lisbon
meridian, the distance between them remaining to be navigated
is virtually annihilated.

The general outline is not unlike that of the Genoese map
of 1457; it is also evident that later cartographers, e.g. Contarini
and Waldseemüller drew on a source common to Behaim for the

features of the Indian Ocean and eastern Asia. We are justified
in assuming on these and other grounds that Behaim had not
gone directly to the authorities he quotes, but had merely
amended an existing world map. No special knowledge of
Conti's narrative is shown, but a certain Bartolomeo Fioren-
tino, not otherwise known, is quoted on the spice trade routes
to Europe. South-east Asia is represented as a long peninsula
extending southwards and somewhat westwards beyond the
Tropic of Capricorn. This feature is a remnant of Ptolemy's
geography, evolved when the Indian Sea was opened to the
surrounding ocean. The placing of Madagascar and Zanzibar
approximately midway between this peninsula and the Cape
must be another feature of some antiquity. Fra Mauro displays
far more up-to-date knowledge of this area.

The new knowledge displayed is confined to Africa, or
rather to the western coast for the names on the east coast, save
for those taken from Ptolemy, are fanciful. The main features
of the west coast are more or less recognizable, though Cape
Verde is greatly over-emphasized. To Cape Formoso, on the
Guinea coast (true position 4° 12' N., 6° 11' E.) the nomenclature differs little from contemporary usage. Beyond it,
though a good deal can be paralleled in the two other contemporary sources, Soligo and Martellus, there are elements
peculiar to Behaim, e.g. the 'Rio de Behemo', near Cape
Formoso, and the 'Insule Martini', identified by Ravenstein
with Anobom, with others of a less personal character. The
coast swings abruptly to the east at 'Monte negro', placed
by Behaim in 38° South latitude. This is the point reached by
Cao in 1483, and its true position is 15° 40' South. A Portuguese
standard marks the spot. On the eastward trending coast, there
are names which seem to be related to those bestowed by Diaz,
and the sea is named 'oceanus maris asperi meridionalis', a
phrase which doubtless refers to the storms encountered by
him. Owing to the exaggeration of the latitudes, 'Monte
negro' falls fairly near the position which the Cape of Good
Hope should occupy. It is noticeable that the Soligo chart
ends in 14° S. which is near the limit of Behaim's detailed
knowledge. We might conclude therefore that Behaim's
contribution was to reproduce this coast from a similar chart,

and to add some gleanings from the Diaz voyage round the Cape. The two personal names are not to be found on any other map: in conjunction with the attempt made to associate Behaim's own voyage with the discovery of the Cape, we are justified in assuming that this portion of the globe at least was designed in a spirit of self-glorification. It seems doubtful if Behaim had sailed much further than the Guinea Coast.

Note: Since the above was written, R. A. Skelton has recorded the discovery of a printed wall-map of the fifteenth century which, with the exception of Behaim's globe, is the only non-Ptolemaic world map of the fifteenth century to be graduated in longitude. He states 'This printed map, or one similar to it, undoubtedly served as the prototype for Behaim's globe', thus supporting the hypothesis in the text that Behaim had amended an existing world map. (R. A. Skelton, The cartographic record of the discovery of America, *Congresso internac. hist. dos descobrimentos,* Lisboa, 1961, vol. 2, pp. 22.)

REFERENCES

ALMAGIA, R., ed., Monumenta Cartographica Vaticana, vol. 1, p. 32 ff., Città del Vaticano, 1944.
——, I mappamondi di Enrico Martellus. (*La Bibliofilia,* Firenze, 42, 1940, 289–311.)
CRONE, G. R., Fra Mauro's representation of the Indian Ocean and the Eastern islands. (*Studi Colombiani, Genova,* 3, 1952, 57–64.)
——, Martin Behaim, navigator and cosmographer; historical personage or figure of the imagination. (*Actas Congresso internac. hist. dos descobrimentos,* vol. 2, Lisboa, 1960.)
FISCHER, T., Sammlung mittelalterlicher Welt- und Seekarten. Venice, 1886.
KIMBLE, G. H., Geography in the Middle Ages, 1938.
KRETSCHMER, K., Eine neue mittelalterliche Weltkarte der Vatikanischen Bibliothek. (*Zeits. Ges. f. Erdkunde Berlin,* 26, 1891, 371.)
RAVENSTEIN, E. G., Martin Behaim; his life and his globe, 1908.
WINTER, H., Martin Behaim: Geschichte und Legende. (*Die Erde,* 90, 1959, 359–362.)
LEPORACE, T. Gasparini, and R. Almagia, Il mappamondo di Fra Mauro. (Introd. to colour facsimile, Venice, 1956.)
ZURLA, P., Il mappamondo di Fra Mauro. Venice, 1806.

CHAPTER V

THE REVIVAL OF PTOLEMY

IN discussing the latest forms of the medieval *mappa mundi*, we have had occasion to refer to the spreading knowledge of the maps accompanying the *Geography* of Claudius Ptolemæus in the fifteenth century, and possibly earlier. We may now examine the circumstances in which copies of the text and maps became available in western Europe, first in manuscript and later with engraved maps in printed volumes.

The earliest surviving manuscripts of Ptolemy's geographical treatise, in Greek, date from the end of the twelfth or the early thirteenth century. Of these there are two versions, the 'A' recension accompanied by twenty-seven maps, and the 'B', with sixty-four. A copy of the 'A' recension was obtained from Constantinople in 1400 by the Florentine patron of letters, Palla Strozzi, who persuaded Manuel Chrysolorus to translate the text into Latin. Chrysolorus, the founder of Greek studies in Italy, was unable to carry through the task, and it was then undertaken by his pupil, Jacopo Angelus of Scarparia, who completed it about 1406. His translation met with criticism, but, corrected and emended by a succession of editors, it formed the basis of all the printed texts for a century. It was first printed, without maps, at Vicenza in 1475. The maps were redrawn and their legends translated from the twenty-seven maps of the 'A' recension in the first decade of the century by two Florentines, Francesco di Lappaccino and Dominico di Boninsegni: the 'B' recension was never made available in translation to western Europe, though details from the maps were on occasion inserted in the others.

The original manuscript of Angelus' translation and the first maps of the Latin version have not survived, but manuscripts are extant from the third decade of the century, for example, that prepared under the direction of Cardinal Guillaume Fillestre in 1427 (known as the Nancy codex). This codex contains in addition a map of the northern regions based largely upon that of Claudius Clavus, on which 'Engroen-landt' is depicted, and also a list of geographical positions. Fifteenth-century cosmographers like Fra Mauro did not accept Ptolemy's views uncritically, and it became the practice to add a number of contemporary maps to the MSS. to provide a basis for comparison.

The MSS. composed by the Florentine, Pietro del Massajo, are particularly notable for these supplementary maps. The earliest,[1] which must have been written before 1458, contains the twenty-seven Ptolemaic maps, "cum additione provinciārum noviter repertarum et alia nonulla". The seven maps of the 'provinces' comprise Spain, France, Italy, Etruria, the Pelopo-nessus, Candia, and Egypt with Aethiopia; the 'others' are nine town plans, including Rome and Alexandria. The origins of these 'modern' maps in some cases go far back into the four-teenth century, and appear to be linked to the early marine charts. The earliest prototype is a map of Italy which is found with a manuscript of Fra Paolino's 'Cronaca'. Paolino was a contemporary and friend of Marino Sanudo, and it was to him that Sanudo's 'Secreta fidelium crucis' was referred for examina-tion by the Pope. The map, which was not drawn by Paolino, has some affinities with those of Pietro Vesconte. The outline and coastal names were undoubtedly derived from con-temporary marine charts; and an attempt has been made to combine with these a representation of the orography of the peninsula. The source of the latter feature is as yet unascer-tained. In the course of time improvements were gradually introduced; in one type the orientation of the peninsula is more accurate, in another, the representation of its southern extremity is less constricted. Massajo's 'modern' map of Italy has the improved orientation, and additional details. No prototype of his map of Spain has yet been found but its evolution was probably on similar lines. The map of Egypt is particularly interesting, as it includes quite detailed and accurate itineraries in Abyssinia. Other codices include a map of the Holy Land which, it is scarcely open to doubt, derives ultimately from that included in the Sanudo atlases.

[1] Bibl. nat. Paris, Cod. lat. 4802.

These maps are also notable for the method of representing relief. The highlands are marked off from the lowlands; and their surface filled in by solid colour: though this method tends to represent all mountains as plateaus, it is possible to see in the dividing line between upland and lowland and in this use of colour the prototype of form lines and layer colouring. There also appears to be an attempt at oblique hill shading.

Before leaving these 'modern' manuscript maps, we may note that for Italy the two improvements mentioned above—in orientation and in the southern configuration—are combined for the first time on a map in another Ptolemy codex. This map in the Laurenziana, drawn about 1460, is important as either it, or a near version, was followed by Berlinghieri, and later by the editors of the early sixteenth-century Rome editions. It is an improvement on Ptolemy's outline and is more correctly oriented—marine charts, and rather ancient ones at that, having been used for this purpose.

More important as a producer of these manuscript atlases was Dominus Nicholaus Germanus. Very few details of his life are known with certainty, and his career has given rise to much surmise. He was undoubtedly in Florence and Ferrara around the period 1464 to 1471. Florence was then a centre of cosmographical studies, and Nicholaus was known to its leading scholars. He seems to have attracted attention by his presentation of a magnificently illuminated manuscript of the 'Geography' to Borso d'Este in 1466. In all, Nicholaus was responsible for twelve MS. copies of the 'Geography'. These fall into three main groups, two of which formed the basis of printed editions. Nicholaus claims several improvements for his versions: the maps redrawn in a smaller and more convenient size; the employment of a new projection (the 'trapezoidal'); the correction of the outlines of the various countries; and the addition of new maps. He undoubtedly made alterations but they were not all improvements, nor innovations devised by himself. The manuscript maps by Nicholaus were the basis of the first printed edition of the 'Geography', Bologna, 1477, and of the Rome edition of 1478: they have therefore an important bearing on the form in which Ptolemy's data were disseminated, through the recently invented printing press and the technique of engraving on copper plates.

Also at work in Florence during these years was Francesco Berlinghieri, who prepared a rhyming version of the 'Geography' and accompanied it with an important set of maps, including a number of modern maps much superior to those of Nicholaus Germanus, being related to the Massajo and Laurenziana types. The first edition was published at Florence in 1482.

Finally there was one other copyist engaged on the 'Geography', Henricus Martellus. A splendid MS. of his is preserved in the National Library at Florence; it contains thirteen modern maps, but is probably later than the earliest printed editions. The map of France and northern Italy is particularly striking. The Alps are elaborately drawn in an 'oyster shell' design, outlined and ribbed in dark brown with a lighter brown and white body. Some summits have flat green tops with small tree symbols. Martellus, who was also the copyist of an important atlas now in the British Museum, was of German origin, but nothing more is known of him.

Thus in the mid-fifteenth century four cartographers were engaged in multiplying copies of the 'Geography' and its maps: P. del Massajo, c. 1458-72; Nicholaus Germanus, 1464-71; Francesco Berlinghieri, and Henricus Martellus, about 1480. It is significant that the first three had connexions with Florence.

The first printed edition of the 'Geography', without maps, was published at Vicenza in 1475, but probably before that date experiments were already being made in engraving maps on metal plates, from which large numbers could be printed. The lead in this work was taken by Conrad Sweynheym in Rome, and his labours finally bore fruit in the magnificent Rome edition of 1478. It however was anticipated by the Bologna edition of 1477. (The title page is erroneously dated 1462.) The maps for this were drawn by Taddeo Crivelli, an accomplished miniaturist and draughtsman, who had been attracted from Ferrara to Giovanni Bentivoglio's court at Bologna. Crivelli was no doubt aware of the acclaim which Nicholaus Germanus had won by the presentation of his illuminated codex to Borso d'Este, and this may have prompted him to propose to Bentivoglio, eager to show himself a patron of learning, the printing of the 'Geography'. The venture was certainly undertaken in a competitive spirit, for it was hurried on to forestall the Roman edition, and it has been suggested that one of Sweynheym's workmen was enticed away from Rome to reveal his technique to the Bologna printers. The manuscript which was used was closely related to one by Nicholaus Germanus, but owing to hasty production, the edition was unsatisfactory. The text is marred by misprints, and the maps are poorly executed, with numerous errors and omissions, and much evidence of inexpertness and haste. Its shortcomings were realized by the publishers, and during the next two years the plates were improved and new versions issued. There is little to be said for this edition; it is certainly the first to contain engraved maps—but otherwise Crivelli showed himself to be a better artist than cartographer, despite the help of two astrologers.

This Bologna edition contains twenty-six ancient maps; they are drawn on the original conical projection, with degrees of longitude and latitude indicated in the margins, and also the climates.

The Rome edition of the 'Geography' finally appeared in 1478, one year after the Bologna edition. The text was edited by Domitius Calderinus, probably using the Ebner Codex of Nicholaus Germanus. The maps were engraved on copper by Conrad Sweynheym, and are very finely executed. ¡The outlines are sharp, and there is a pleasing absence of unnecessary detail. The names are in a style based upon the lettering on the Trajan column, and set a high standard for later map engravers. Mountain ranges are drawn in profile, rather in the style of 'mole hills'. Given the magnitude of the task, and the experimental stage of the art of engraving, the atlas is an extremely fine production. The maps are the twenty-seven ancient ones of the 'A' recension, on the rectangular projection: degrees of latitude and longitude are marked in the margins, and also the length of the longest days.

The first printed work to include 'modern' maps with Ptolemy's maps is strictly speaking not an edition of the

'Geography', but Berlinghieri's metrical version of that work, printed at Florence in 1482, is of sufficient importance to be noted with this series. The maps, boldly engraved on copper, are thirty-one in number, the four additional ones being "Hispania Novella", "Gallia Novella", "Novella Italia", and "Palestina moderna". These new maps are on the original rectangular projection; latitude and longitude are not indicated in any way, nor have they scales. Their outlines are clearly derived from the Laurenziana codex or a very close source. The influence of the marine charts is clearly visible in the style of the coastlines, with numerous semi-circular bays and conspicuous headlands. The representation of relief is also very similar to the Laurenziana manuscript. The names on these modern maps are in the current popular forms. They are certainly the most accurate maps to have been printed in the fifteenth century, and it was unfortunate that they were overshadowed for the time by the so-called modern maps of Nicholaus Germanus in the Ulm editions, and to some extent by the Ptolemy maps themselves. The Berlinghieri maps were reprinted again, probably after 1510, and they also had some influence upon the Rome editions of 1507 and 1508.

The next edition was edited by Nicholaus Germanus himself, and printed at Ulm in 1482. Thus in the period 1477-82, four editions with maps had appeared, three in Italy and one in Germany. As one thousand copies of the Bologna edition were printed, and the other editions were probably of a similar size, Ptolemy's ideas received wide diffusion just at the moment when they were about to be, to a large extent, proved erroneous. There are thirty-two wood-cut maps in the Ulm Ptolemy, a 'modern' map of Scandinavia, based to some extent on that by Claudius Clavus, having been added to the four new to the Berlinghieri edition. The Ptolemaic world-map, for the first time in a printed work, has been amended, the north-west sector being drawn to accord with new details of Scandinavia. The maps, original and modern, have all been redrawn on the 'trapezoidal' projection which Nicholaus claims for his own. It may be regarded as a crude conical projection, the meridians radiating from the Pole, and the parallels being drawn at right angles to the central

C*

meridian. On the modern maps there are no indications of latitude and longitude, though the length of the longest day is noted at intervals in the margin. As these figures are based on latitude they afford some indication of position—but the reluctance, or perhaps the inability, to show it more accurately is curious: it is not until the Rome editions of 1507 and 1508 that this defect is remedied. In drawing the new maps Nicholaus adopted a very conservative attitude; for all practical purposes he accepted the outlines of Ptolemy, modified in some details by the later maps mentioned above, and attempted to fit the new detail in this frame, with, as might be expected, very unsatisfactory results.

On the whole this edition can only have had a retrogressive effect on the development of cartography. It seems, however, to have met with a good reception in Germany, for within four years a second edition appeared at Ulm (1486), with the same maps and the text enlarged by a dissertation. In 1490, a second edition of the Rome version of 1478 appeared, with the twenty-seven maps printed from the same plates. There was then an interval of seventeen years before another edition was issued. This coincided with the great epoch of maritime expansion, and naturally, until adequate details of the new discoveries became available, there was little incentive to embark on a new edition.

The third Rome edition appeared in 1507, edited by Marcus Beneventanus and Johannes Cotta. The twenty-seven ancient maps are from the plates of the earlier editions, and to these were added six new maps, engraved in a similar style. Five of these had appeared in slightly different forms in other editions, but the sixth was of greater interest. This was a map of Central Europe (Polonie, Hungarie, Boemie . . .), by Cardinal Nicholaus Cusanus. A manuscript copy is in the Laurenziana codex, and it had apparently been intended to include it in one of the earlier Rome editions; a plate was engraved but not used for this purpose, though the map was in circulation separately about 1491. The other 'tabulae modernae' are derived partly from the Ulm editions (northern Europe, France, and the Holy Land—the first two on the trapezoidal projection) and partly from Berlinghieri (Italy, a close copy, and Spain, on the rectangular projection). For the first time, the new maps have borders graduated for latitude and longitude, and numbered in degrees. It is significant that a legend on the modern map of Italy states that the measure of the degree of longitude does not follow Ptolemy, but is shown "according to the style of the nautical charts". This appears to mean that the map is drawn on a plane projection—that is that no allowance is made for the convergence of the meridians, for a degree of longitude is made to equal a degree of latitude (very nearly).

The following year these plates were used again for another edition of the 'Geography', enlarged by the addition of a short treatise on the new world by Beneventanus, and—of much greater importance—Johan Ruysch's world map. This was the first map in an edition of Ptolemy to show any part of the new world.

Three years later an edition was published at Venice by Bernardus Sylvanus which made a further break with tradition. The twenty-seven maps were re-engraved on wood with many names stamped in red: they have 'modern' outlines. with the classical nomenclature; and there are thus no strictly Ptolemaic maps in this edition. In this way, therefore, a printed map of the British Isles, other than that by Ptolemy, was for the first time placed in circulation. It was not a very accurate one, being based on the portolan chart by Petrus Roselli, and a few names from it are also included. The whole is crudely drawn, London, for example, being shown well to the south of the Thames. The world map is on a heart-shaped projection, and is brought in line with contemporary knowledge; Hispanola, Cuba and a part of South America are inserted, and the complete coastline of Africa—but in the east Ptolemy's outline is retained.

The peak of Ptolemy's influence on cartography was reached with the edition of the 'Geography' published at Strasbourg in 1513. This is put forward as the work of Jakob Eszler and Georg Ubelin, but the maps are generally accepted as the work of Martin Waldseemüller (1470-1518) of St. Dié in Loraine, though conclusive proof is lacking. At St. Dié, Waldseemüller was a member of the scholarly circle patronized by the Duke, René II. The maps form, with his other works—the 'Cosmographiæ introductio', a globe, and two world maps of 1507 and 1516[1]—a related body of old and new geography, anticipating the scheme of Gerhard Mercator. The atlas contains forty-seven woodcut maps, of which eleven may be considered as new. These include a world chart which is a crude version of his elaborate 'Carta marina' of 1516, based in turn on Canerio's chart; a 'Tabula terre nove', one of the earliest separate maps of the American continent; a map of Switzerland based on a manuscript map of Konrad Dürst of 1496; and a 'Tabula moderna Lotharingiae'. The latter is of interest as an early example of printing in colour, and for its attempt at depicting the landforms of the region.[2]

This edition was reprinted from the same blocks in 1520, and two years later Laurent Fries put out another, with somewhat different maps on smaller scales, but also attributed to Waldseemüller. Though many new maps were pouring from the presses, the interest in Ptolemy did not die out completely in the sixteenth century: of those editions which preceded Mercator's, perhaps the most important were those by Sebastian Münster (Basel, 1540) and by Jacopo Gastaldi (Venice, 1548), the latter, a small octavo, containing sixty engraved maps, generally based on Münster's, but with considerable additions. Not long after, these composite collections of old and new geography were to be superseded by the modern atlases of Ortelius and Mercator.

[1]For these maps, see p. 99 below.
[2]Taylor, E. G. R., A regional map of the early sixteenth century. (*Geogr. Journ.*, 71, 1928, 474.)

REFERENCES

ALMAGIÀ, R., Monumenta Italiae Cartographica, Firenze, 1929.
DURAND, W., The Vienna Klosterneuburg map corpus. Leiden, 1952.
FISCHER, J. and VON WIESER, F., The oldest map with the name America, etc. Innsbruck, 1903.
GALLOIS, L., Les géographes allemands de la Renaissance. Paris, 1890.
LYNAM, E., The first engraved atlas of the world; the *Cosmographia* of Claudius Ptolemaeus, Bologna, 1477. Jenkintown, 1941.
STEVENS, H. N., Ptolemy's 'Geography': a brief account of all the printed editions to 1730. 2nd ed., 1908.

CHAPTER VI

THE CARTOGRAPHY OF THE GREAT DISCOVERIES

THE second great contribution to the revival of cartography was made by the leaders of overseas expansion; the seamen of many nations—Italian, Portuguese, Spanish, French, Dutch and English—who in little more than a century opened up the oceans of the world, with the partial exception of the Pacific, and provided the chart makers with the data for the maps of their coastlines. The outstanding stages in this progress are: the rounding of the southern promontory of Africa by Bernal Diaz in 1487; the landfall of Columbus in the West Indies in 1493; the attainment of India by Vasco da Gama in 1498; the discovery of Brazil by Cabral in 1500; the capture of Malacca by Alfonso d'Albuquerque in 1511; the arrival of the first Portuguese in the Moluccas in the following year, and the circumnavigation of the globe by Magellan's expedition.

In order to judge the standard of accuracy of these charts, we must glance rapidly at the methods of navigation practised by these pioneer seamen.

At the outset of their African voyages, the Portuguese pilots followed the same methods of navigation as the sea-faring peoples of the Mediterranean. From the marine charts they ascertained the direction, or rhumb, of the proposed voyage, and also its distance. With the aid of the mariner's compass and primitive methods of determining the vessel's speed, they tried to keep as close as possible to this track, estimating their position daily. In the Mediterranean, voyages were largely but not exclusively a matter of coastwise sailing, so that much reliance was also placed on acquired knowledge of local winds and currents and on the ability to recognize prominent coastal landmarks, a bold headland, a group of islets, or a distinctively shaped mountain. Pilots in the Mediterranean therefore rarely troubled to determine their latitude, partly also because the latitudinal range was relatively small, and the degree of accuracy of their observations not high.

When the Portuguese embarked upon the waters of the Atlantic and made their way southwards along the African coasts, they encountered different conditions. There was no body of traditional seafarer's lore to draw upon, as regards winds and currents; familiar landmarks were lacking on the coasts, which were often characterless for considerable stretches, and fringed by unseen hazards—moreover a hostile population discouraged unnecessary approaches. Added to these was the possibility of being driven off course into the ocean. They were also ranging through many degrees of latitude. In these circumstances the pilots turned to the determination of latitude, at first by observing the altitude of the Pole Star. Later, as the vessels pressed further southwards, and the Pole Star sank lower in the sky, latitude was obtained from the midday altitude of the sun with the help of tables of declination. These observations were made with the astrolabe, gradually simplified from the landsman's type, and with the quadrant, a less cumbrous instrument.

Since the Pole Star does not coincide with the celestial pole, it was necessary to apply a correction to its observed altitude in order to obtain the latitude. The correction depended upon the time of the observation, which could be obtained from the position of Ursa Major in its orbit round the Pole. A set of simple instructions, known as 'The Regiment of the North', was therefore drawn up, which gave the correction to be applied for certain positions of the 'Guards'.

For the corrections to be applied to the sun's midday altitude, a primitive table had probably been worked out by 1456. Later José Visinho, utilizing the work of the Jewish astronomer Abraham Zacuto, calculated a table for each day of the leap year March 1483-February 1484, and this was used by Bartholomeu Dias in his famous voyage. Later still Zacuto assisted in preparing a perpetual almanac for the voyage of Vasco da Gama.

It will be seen that these scientific aids were provided relatively late in the fifteenth century; the first recorded use of the quadrant at sea dates from 1460. It is not until the early years of the sixteenth century that scales of latitude are found on marine charts, so that until then the charts which recorded the Portuguese advance along the African coasts continued to display the features of the Mediterranean portolan charts. Since the coasts they were charting ran for the most part in a southerly direction, this was not at first a source of great difficulty, though the influence of magnetic variation was not appreciated. It had been observed on land, but its calculation at sea was not seriously undertaken until the next century. When it became necessary to chart accurately a number of points, situated for example on each side of the Atlantic and extending through many degrees of latitude, the neglect to take account of the convergence of the meridians rendered the old type of chart extremely inaccurate. The task before the pilots at this stage was still largely to set down distance and direction as accurately as possible, but when we examine the charts which have survived, for example, Andrea Bianco's at the beginning of the period (1448) and Graciosa Benincasa's chart of the discoveries some distance beyond Sierra Leone (1468), we have to conclude either that they were very carelessly assembled from sectional charts, or that accuracy of distances was forfeited for some other purpose. Benincasa's chart portrays the coast on a scale which increases steadily southwards, the scale of the most southerly portion being nearly four times that of the northern. Now the northern portion of the coastline is featureless in contrast to the estuaries and islands further south, and it is possible that the diversified coast was deliberately drawn on a larger scale. A similar variation in scale and emphasis on prominent features are characteristic also of Bianco's chart. In the brief reference above to methods of navigation, the extent to which landmarks were employed in coastwise sailing has been pointed out, and by this enlargement of scale features were made more easily recognizable. The descriptive names used on the charts served the same purpose.

When the latitudes of a number of places on the African coast had been fixed, it became less necessary to emphasize particular stretches in this way, for the navigators were not tied to coastwise sailing. In rounding the Cape, for example, it was the practice to sail southwards as rapidly as possible to the necessary parallel of latitude and then to turn eastwards, keeping as close as possible to the parallel ('running down the easting'). If through navigational errors the African coast were sighted to the north of the Cape, it was a simple matter to coast southwards. To assist in determining their latitude, navigators were provided with tables, known as 'Rules of the Leagues', which simply stated the number of leagues it was necessary to sail on various bearings in order to make good one degree of latitude, north or south (i.e. the hypotenuse of a right-angled triangle, of which one other side equals 1° or 70 miles).

Since the astronomical determination of longitude was a process of great complexity before the invention of accurate time-keepers, all east-west distances depended upon dead-reckoning alone. From the courses and distances run, it was possible by the application of the 'Rules' to calculate each day's sailing and ultimately the total voyage. Given the length of a degree of longitude at various latitudes, it was possible to arrive at an approximate figure for the difference of longitude. It is necessary to keep these considerations in mind when discussing the accuracy of the charts which record the great discoveries.

All these maritime achievements, in the east and in the west, were accomplished within thirty-five years, and it might be expected that the cartographic output for this period would be large. Actually, despite the momentous events to be recorded it is not great—or to be more accurate, the material which has survived is relatively slight.

No original chart from the period 1487 to 1500 has been preserved. The nearest is the copy of a chart of the western African coasts to the neighbourhood of the Cape of Good Hope contained in the collection generally known by the name of the copyist, Soligo, which was probably made about 1490 (B.M. Egerton 73). The representation of Africa in the globe of Martin Behaim and the map of Henricus Martellus may possibly be based on contemporary charts at second or third hand—but otherwise this decade is cartographically a blank. If the period is extended to 1510, the number of survivals is still relatively small; the more important are the world charts, or

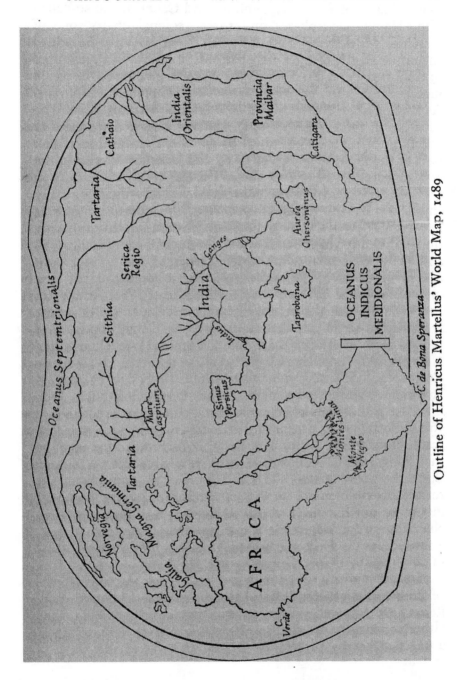

Outline of Henricus Martellus' World Map, 1489

planispheres, of La Cosa, Cantino, and Canerio; the so-called King-Hamy planisphere; and three regional charts, one of which is certainly by Pedro Reinel. To these may be added the crude sketch of the northern coast of Hispaniola attributed to Columbus, and the diagrammatic representation of the world incorporating the new discoveries by Bartholomeo Columbus. As will be seen, the secondary material also is not large. The time-lag in the appearance of maps of the new discoveries to satisfy public interest is shown by the fact that no map of any part of the New World or of the Portuguese discoveries in the east appeared in a Ptolemy atlas until 1507.

As it is known from contemporary records that many charts were made during this period, the question arises as to why so few have survived. The reason for this is partly that in the early years charts were in high demand by navigators, and would consequently be widely dispersed and rapidly worn out or lost. As to Portuguese charts, a large number was undoubtedly lost in the Lisbon earthquake of 1755. As a widespread interest in the discoveries only developed slowly, there is also little secondary material. Vasco da Gama's opening of the sea-route to India in 1498 and Vespucci's account of the New World (popularized by Waldseemüller's 'Introductio Cosmographiae') were the events which really caught popular attention. The first collection of voyages, which brought both the East and West Indian voyages together, appeared only in 1506. Portuguese historians have argued that this was the result of an official policy of secrecy. It is recorded for example that King John II imposed a ban on the circulation of charts. Since however pilots and cartographers passed from the service of one monarch to that of another, apparently without incurring much odium, it must have been difficult to keep charts secret for long, and we shall see that, after 1500 at least, a few copies of charts recording the discoveries were available in Italy.

In keeping with the general paucity of cartographic material relating to the earliest years of the discoveries, only two small items have survived which can with any certainty be ascribed either to Christopher Columbus or to his brother Bartolomeo, a chart maker by profession. In the archives of the Duke of

Alba in Madrid, there is a hasty outline sketch of the north and north-western coastline of Cuba, on which occurs the name 'nativida', for La Natividad, the first settlement in the New World, which Columbus had founded on his first voyage. This is ascribed to Christopher. The second consists of three marginal sketches in a copy of Columbus' letter of July 1503 describing his fourth voyage, preserved in the National Library, Florence. From this we learn that a regular survey of the Central American coast was carried out by Bartolomeo. The sketch maps, ascribed to him, form an outline of the world between the tropics, and are of particular interest as they illustrate very clearly Columbus' ideas on the relationship of his discoveries to south-eastern Asia. The north coast of South America is prolonged westward before it joins that of Central America, and the latter is joined to the Asian coastline of Ptolemy in the neighbourhood of Cattigara. This synthesis required the placing of Central America 120° of longitude to the west of Cape Verde!

Some years ago Charles de la Roncière drew attention to a circular world map in the Bibliothèque National, Paris, which he argued had been prepared at Columbus' direction for presentation to the Spanish sovereigns. La Roncière's arguments have been strongly controverted; in any case, the map ante-dates the first voyage, and does not throw any significant light upon Columbus' aims.

There is some doubt as to which is the earliest extant chart to show any of the discoveries in the New World. This is either a chart by La Cosa, or an anonymous one, known by the name Cantino, which can be definitely dated 1502. The La Cosa chart bears the date 1500, but this has been challenged. We may accept it for the chart as a whole, though some additions were probably made to it later, for G. E. Nunn's arguments for dating it c. 1508 are not entirely convincing.

Juan La Cosa, an expert Biscayan navigator (not the owner of the *Santa Maria*), accompanied Columbus on his second voyage. He later made further voyages to the American continent, and is known to have drawn several other charts since lost. The map, 180 × 96 cms., somewhat crudely drawn on parchment, has suffered considerable damage. On the western

margin, below a drawing of St. Christopher in the neck of
the skin, is the inscription "Juan de la Cosa la fizo en el puerto
de s: maria en año de 1500". The chart is in the style of earlier
marine charts, with compass roses and direction lines. The
scale is given by a line of dots, unnumbered and unexplained;
the distance between the points however is apparently intended to represent fifty miles. The northern tropic and the
equator are drawn, but degrees of latitude or longitude are not
indicated. In the west are the discoveries of Cabot in the north
and of Columbus and the Spaniards in the West Indies and
along the north-eastern coasts of South America. The Bahamas
group is shown with some accuracy but necessarily on a small
scale. It includes the island Guanahani, Columbus' first landfall,
alternatively known as San Salvador and now identified with
Watling Island. No special emphasis is given to this memorable
locality. Off the South American coast is a large "island
discovered by the Portuguese", representing Cabral's discovery of Brazil in 1500. The chart-maker appears to have
considered the American coastline to be continuous from
north to south, but this cannot be asserted with certainty,
as the Central American area is hidden by the drawing of St.
Christopher.

The eastern margin of the map cuts down the continent of
Asia beyond the 'Ganges', so that the coastline is not shown.
The most conspicuous feature in this quarter is the triangular
island of 'Trapobana'.

In latitude the map extends from the Scandinavian peninsula to the southwards of the African continent. The African
coast as far as the Cape of Good Hope is represented with fair
accuracy, from Portuguese sources. The eastern coast however
seems to be entirely imaginary. In the Indian Sea, almost in
the centre, are two large islands, 'Zanabar' and 'Madagascoa',
as on Behaim's globe. The sole indication of da Gama's voyage
is the inscription "Tierra descubierta por el Rey don Manuel
de Portugal" on the south coast of Asia; the outline of the
coast however is no improvement on that of the Catalan map of
1375.

The map in fact has every appearance of having been put
together from at least two sections: the western portion com-

prising the American discoveries and perhaps the West African coasts having been joined to a portion of a world map resembling those of fifty years earlier which display the influence of Ptolemy. If we use the distance between the tropic and the equator to obtain a scale of degrees, and apply this to the map, we find that in the western section, though there are discrepancies, the general picture is not wildly inaccurate. The newly discovered lands are placed in fair relationship to those of western Europe. The longitudinal difference between ,the Iberian coast and Hispaniola is apparently about 62°, instead of 59°, and between the African coast and the north-east coast of South America approximately 16°, instead of 17¾°. For a reason which has never been satisfactorily explained, Hispaniola and Cuba are placed well to the north of the tropic; the north coast of Cuba being shown in approximately 36° N., some 12° too far north. Whatever the reason for this, it would appear that the Central and South American portion is on a larger scale than the rest of the map.

The representation of Africa is distorted by the excessive length of the Mediterranean. The general shape of the western coastline is good, though, in relation to the west-east extent of the Gulf of Guinea coast, the coastline southwards to the Cape is too short. This was a characteristic of early Portuguese charts of this region: owing to adverse sailing conditions, it was usual to underestimate distances run.[1]

Much attention has been attracted to the representation of the north-eastern coastline of America. The principal features are (1) a prominent cape, 'Cavo da Yngleterra', about 1,300 miles from south-west Ireland, and approximately in the same latitude; (2) to the west of this Cape, an extent of coastline, running about due west for approximately 1,200 miles: a number of features along this coast are named, and this is the only portion of the North American coastline on which names occur; (3) beyond this coast, a stretch without names continues for another 700 miles, forms a bay, 'Mar descubierta por Yngleses', and then turns southwards.

The 'Cavo da Yngleterra' is shown in about 56° N.

[1]See 'Esmeraldo de situ orbis' by Duarte Pacheco Pereira, ed. G. H. T. Kimble. (*Hakluyt Soc.*, ser. ii, vol. 79, 1936.)

latitude. Since, however, the latitudes of many places in Europe are out by several degrees (Land's End, for example, is shown $4\frac{1}{2}°$ too far north) the Cavo may be assumed to be not further north than 51° 30' N., which would put it in the neighbourhood of Belle Isle Strait. On the other hand, the 1,200 miles of explored coastline is in all probability southern Newfoundland or Nova Scotia, so that the Cavo de Yngleterra must have lain further south, and Cape Race at once suggests itself, but as nothing more than a possibility. J. A. Williamson, however, who credits this charting to the Cabots in 1497–8, believes that the Cavo was Cape Breton, while G. E. Nunn identifies it with Cape Farewell in Greenland.

The earliest Portuguese example of these New World charts is that known as the Cantino chart. It owes its name to the fact that it was procured for the Duke of Ferrara, Hercules d'Este, by one Alberto Cantino. The Portuguese king had placed an embargo on the provision of charts showing the new discoveries, and Cantino had obtained this clandestinely to satisfy the curiosity of the Duke, anxious at the threat to the Italian share in the spice trade. As correspondence concerning the transaction has survived, we know that the chart was received by the Duke in November 1502, and that it embodied discoveries as late as the summer of that year. The chart is clearly the work of a Portuguese cartographer; at a later period apparently some amendment has been made to the Brazilian portion, and half a dozen Italianized names written in. The title given to it suggests that the main interest of the draughtsman was in the western discoveries: "Marine chart of the islands recently discovered in the parts of the Indies."

The chart is large, so that the coasts are shown in considerable detail, and names are numerous. The Equator and tropics are drawn in, but there is no graduated scale of latitudes. From west to east it extends from Cuba to the eastern coast of Asia. The Tordesillas demarcation line between the Spanish and Portuguese spheres is inserted, and the Portuguese discoveries in the north-west are made to lie just on the Portuguese side of the line.

The African continent is shown for the first time with something closely approaching its correct outline: on the east coast the names of Soffala, Mozambique, Kilwa, and Melinde occur, and the island of Madagascar is inserted but not named. The Indian sub-continent is drawn as a sharply tapering triangle, on the western coast of which are names—e.g. Cambaya, Calecut—and legends detailing the wealth of these parts, which were drawn from accounts of Vasco da Gama's voyage. These appear to mark the limit of first-hand knowledge; beyond, the outline must have been inserted largely by report. That this was obtained from native seamen is probable from the circumstance that the term 'pulgada' is used in place of a degree; it equalled about 1° 42′ 50″. The places whose latitudes are given thus are inserted only approximately in their correct positions. East of India is a large gulf and then a southward-stretching peninsula, a relic of the coasts which Ptolemy believed to enclose the Indian Ocean. Near its extremity occurs the name 'Malaqua', and off it the large island of 'Taporbana' (Sumatra). The eastern coast of Asia runs away to the north-east, almost featureless but with a number of names, mostly unidentifiable, on the coast and indications of shoals off shore. Recognizable names are 'Bar Singapur' (Singapore) and 'China cochin'.

The main feature to be noted with regard to Asia is the almost complete abandonment of Ptolemy's conception of the southern coasts, and the great reduction in the longitudinal extent of the continent. The south-eastern coastline of Asia is shown as lying approximately 160° east of the line of demarcation, a figure very close to the truth.

The so-called King-Hamy chart also dated 1502 is interesting as showing the Ptolemaic conceptions of Asia in the process of being fitted to the new discoveries in the west. This chart has many features of the Ptolemy world map in south-east Asia, where 'Malacha' and 'Cattigara' appear together, but the point of importance is that the longitudinal extent eastwards from the demarcation line to the south-east Asian coast is still approximately 220°–230°.

The Cantino chart therefore demonstrates clearly that Portuguese cosmographers had entirely abandoned the Alexandrian's figures, and were already aware that the Spanish discoveries in the west, far from neighbouring on Cipangu and the Asian mainland, were separated from them by an interval of almost half the circumference of the globe. The chart might be said to predict the existence of the Pacific Ocean. The fact that the cartographer has a legend on the discoveries in the north-east American shores stating that they were thought to be part of Asia does not controvert this. For the Portuguese, theoretical and practical considerations happily coincided in this instance; when the question of sovereignty over the Moluccas arose, it was to their interest to reduce the longitudinal extent of Asia in order to bring the coveted islands within their sphere.

Another world chart, slightly later than the Cantino, but derived from a very similar source, has also survived. This is a copy made by an Italian draughtsman, Nicolay de Canerio of Genoa, and assigned to the year 1505 or 1506 on the evidence of its portrayal of the Brazilian coast. The interest of this chart lies in the fact that it is the basis of Waldseemüller's wood-cut world map of 1507. In general it is less accurate than the Cantino chart, particularly in its representation of Africa and India, although it places the Cape of Good Hope in the very accurate latitude of 34° S. (for 34° 22′ S.). Off the mainland of north-east Asia is an island 'Chingirina' with the legend "This island is very rich, and they are Christians; thence comes the porcelain to Mallacca. Here there is benzoin, aloes, and musk." It has been suggested that this is a reference to Japan.

These world charts are evidence of the great interest taken in Italy in the Portuguese progress eastwards; had not these copies been demanded by Italian patrons, much valuable cartographic evidence would be lost to us. They further show that much knowledge of the east had filtered through to the Portuguese before they reached Malacca.

In addition to these world charts, there are from the first decade of the sixteenth century a few charts of smaller areas. Three of these are of special interest: a chart of the North Atlantic, c. 1502, signed by Pedro Reinel; a chart of the North and South Atlantic, c. 1506 (generally referred to as Kunstmann III); and a chart of the Indian Ocean of about 1510.

The Pedro Reinel chart, the earliest signed work of a Portuguese cartographer, introduces the feature of the 'oblique meridian'. Off the land of Corte Real in the north-west Atlantic, is a scale of latitude, additional to the main scale, and placed obliquely to it. It has been shown by H. Winter that this is intended to indicate the geographical meridian in this area, and that the angle which it makes with the main meridian is the magnetic variation, in this instance $22\frac{1}{2}°$ W. Since the ordinary pilot would not be equipped to determine the variation, the coasts were laid down by the compass without correction, and the oblique meridian gave the allowance to be made when they were transferred to a graticule of latitude and longitude.[1] The Kunstmann III chart has a scale of latitude divided in degrees; the value of a degree, according to the scale of leagues, is 75 miles, a more accurate value than that usually adopted.

On these and other early Atlantic charts, the outline of Cuba at first resembles that on the La Cosa chart, and the island is placed in a high latitude. About 1506, the curious 'caterpillar' outline is abandoned. They show the progressive exploration in the north-west, the 'Terra Corte Real' (Newfoundland) and 'Terra do Lavrador' (probably Greenland). From the evidence of these charts, one may conclude that the coastline hereabouts on the La Cosa chart very probably should be sought south-west of Cape Race.

The chart of the Indian Ocean, c. 1510, may best be discussed with the charts of the next decade. But before passing to them, something should be said of Pedro and Jorge Reinel, the leading Portuguese cartographers of this epoch, who served the Portuguese crown for many years. Pedro, described as "master of charts and of navigation compasses", was probably the draughtsman of, among others, the important 1518 chart of the Indian Ocean discussed below. During the preparations in Spain for Magellan's voyage of circumnavigation, the Reinels played a somewhat mysterious part. Jorge was at Seville in 1519, and appears to have made a globe and a world map for Magellan's use when arguing his cause with the Spanish king. There he was joined by his father, who also provided the expedition with two maps, which were taken on the voyage. It seems that neither had actually entered the

[1]Winter, H., *Imago Mundi*, vol. 2, 1938, and Taylor, E. G. R., *Ib.*, 3, 1939.

service of Spain, and it has been suggested that they were in Seville partly to discuss the question, which the success of the voyage would raise acutely, of whether the Moluccas lay on the Portuguese or the Spanish side of the line of demarcation. They were then referred to as "pilots of much renown", and five years later, the Emperor Charles V was endeavouring to induce them to enter his service. This attempt failed, and in 1528 they were awarded pensions by the King of Portugal. In 1551, Jorge, who continued to produce charts, was described as "examiner in the science and art of navigation". Later he fell on evil days, for in 1572 he was said to be "sick, old and poor".

The fact that the Moluccas, the principal source of the oriental spice trade, lay near to the Spanish-Portuguese demarcation line in the opposite hemisphere had a stimulating effect upon the study of cosmology and cartography. Both sides were naturally anxious to prove that the islands lay in their sphere, and the issue was sufficiently close, given the means at the disposal of the protagonists, to ensure that the problem was thoroughly discussed with the aid of the latest charts. In the western hemisphere, the line of Tordesillas was the meridian of 46° 37' W. of Greenwich, so that in the eastern hemisphere it fell on the meridian 133° 23' E. As the Moluccas are in approximately 127° 30' E., and the Portuguese sphere lay west of this meridian, the islands were about 6° inside it. Bearing this in mind we may trace the evolution of the cartography of the Indian Ocean and the Eastern Islands from its first blend of ascertained fact and native reports to the completion of relatively accurate charting. It is noticeable that the chart makers hardly indulged at all in conjecture about what was unknown to themselves, or set down the traditional outlines. Their charts are a combination of first-hand knowledge and a restrained use of native information.

The earliest of these Portuguese charts which has survived dates from about 1510. Nothing is now known of the circumstances of its construction or the name of the cartographer. The chart has two scales of leagues and a scale of latitude from 60° S. to 60° N., and is provided with a system of compass roses and direction lines. The representation of the coasts of

Africa and the west and south-east coasts of the Indian peninsula is very fair. Prominent in the Indian Ocean are the Maldive
Islands, running N.W.–S.E., as on the Fra Mauro map. Beyond
south-east India there is a great gap; then in the south-east is
a portion of the southern extremity of the Malay peninsula
with the large island of Taprobana (Sumatra) to the west of
it, between 1° 20′ and 9° 30′ S.

Some of the latitudes shown on the chart are quite accurate,
the Cape of Good Hope is placed in 35° S. (for 34° 20′ S.);
Goa in 15° N. (for 15° 30′ N.) and Cape Comorin in 7° 15′ N.
(for 8° 12′ N.). On the other hand, the Malay peninsula is
brought south to 16° S. (instead of 2° N.), though Sumatra is
only some 5° too far south. The Portuguese were now placing
these eastern islands fairly accurately. The longitudinal extent
of the Indian Ocean along the Equator, from north-east Africa
to Sumatra, is shown as 54° 20′, the actual figure being approx.
52°. The eastern portion of the ocean is, however, contracted
(Maldives-Sumatra is 17°, instead of 22°), while the western
portion, probably under the influence of Ptolemy, is enlarged
(East Africa-Maldives is 37°, instead of 30°). Over the Malay
peninsula the cartographer writes: "Has not been reached
yet."

Within two years of the construction of this chart, the
Portuguese were in possession of a remarkable source of
information, described in a letter from the Viceroy, Albuquerque, to King Manoel. This was no less than a large map
with the names in Javanese, done by a Javanese pilot; it
contained the Cape of Good Hope, Portugal and the land of
Brazil, the Red Sea and the Sea of Persia, the Clove Islands,
the navigation of the Chinese and the Gores,[1] with their
rhumbs and direct routes followed by the ships, and the
hinterland, and how the kingdoms border on each other. In
Albuquerque's words "this was the best thing I have ever
seen". This map was lost in a shipwreck in 1511, but a tracing
of a portion, the most important portion, had been made by

[1] Tomé Pires identified the Gores with the inhabitants of the Liukiu
Islands, in which Formosa was then apparently included. For Pires, Rod-
riguez, and the Javanese map see A. Cortesão, 'The Suma Oriental of Tomé
Pires, etc.' (*Hakluyt Soc.*, Ser. 2, vols. 89 and 90, 1944.) The quotation in the
text is by courtesy of the editor.

Francisco Rodrigues, with the names transliterated, and this
was sent to the King.

> "Your Highness can truly see where the Chinese and
> Gores come from, and the course your ships must take to
> the Clove Islands, and where the gold mines lie, and the
> islands of Java and Banda, of nutmeg and maces, and the
> land of the King of Siam, and also the end of the navigation
> of the Chinese, the direction it takes, and how they do not
> navigate further."

Albuquerque was not slow in following this up, and a small
expedition was despatched which reached Banda in 1512. The
Rodrigues mentioned above was a pilot on this voyage, and the
draughtsman of a series of charts including several of the
south-eastern archipelago and the coasts of eastern Asia. These
charts are assigned by Cortesão to the year 1513. Those of the
archipelago were no doubt partly based on Rodrigues' own
observations, but it may be assumed that they also embody
detail from the Javanese chart. Rodrigues himself did not get
further than Banda. Several features of his charts were long
current in later cartography, e.g. the exaggerated length of the
western coastline of Gilolo (Halmahera). On the other hand,
the more correct notion of the true proportions of the Indian
peninsula was not embodied in charts for some years.

By 1518 these eastern islands are a feature of Portuguese
general charts, for on a chart of the Indian Ocean, preserved
in the British Museum, and ascribed by Cortesão to Reinel,
are depicted Java, Sumbaba and the northern coasts of two
other islands. Further to the east again, is an island group, the
names of which are now illegible, marked by the Portuguese
standard. The question to be settled was the position in
longitude of these islands. The solution can best be followed
on world charts, to which we must now turn.

Sectional charts similar to those discussed above, and many
now lost, were incorporated in the world charts. The most
important of these was undoubtedly the Spanish 'Padron
Real'. This chart, which was the official record of the discoveries, was first made by order of King Ferdinand in 1508.

The duty of revising it as exploration progressed was entrusted to the officials of the Casa da Contratación at Seville. Unfortunately, no authenticated copy has survived, but there are charts by official cartographers, which undoubtedly embody its main features. Owing to the presence of Portuguese chart makers in Spain, much Portuguese work found its way into these charts—in fact our knowledge is largely based on copies by Diego Ribero—and they may be regarded as joint Hispanic-Portuguese productions. Ribero, a Portuguese by birth, was expelled from his native country, and in 1519 was at Seville in contact with the Reinels when preparations were being made for Magellan's voyage. Five years later, described as "our cosmographer and master maker of charts, astrolabes and other navigation instruments", he was a technical adviser to the Spanish representatives at the Conference of Badajoz, when the attempt to negotiate an agreement with Portugal on the ownership of the Moluccas failed, both sides firmly maintaining their claims. Ribero reached a position of considerable eminence in the Spanish service, in which he remained until his death in 1533. By a royal decree of 1526, he was to be provided with all material for a chart and world map portraying all the discoveries, evidently a revision of the 'Padron Real', and the following year he was appointed an examiner of pilots during the absence of Sebastian Cabot on an expedition.

Three world charts similar in type have survived out of all his work, and in view of his official position they may be assumed to be based upon the 'Padron Real'. One, dated 1527, is unsigned, but there are two signed copies, dated 1529. Some comments on the 1529 chart, now in Rome, may fittingly conclude this account of the fundamental Lusitano-Hispanic contribution to the mapping of the world.

Ribero's chart is a landmark in the development of knowledge of the world, comprehending the whole circuit of the globe between the Polar circles, with the East Indian archipelago appearing in both the west and the eastern margins. The placing of the continents in latitude and longitude is on the whole good. The exaggeration of the easterly extent of Asia, however, is still allowed to stand, Canton being placed about 20° long. too far to the east. The area around Canton,

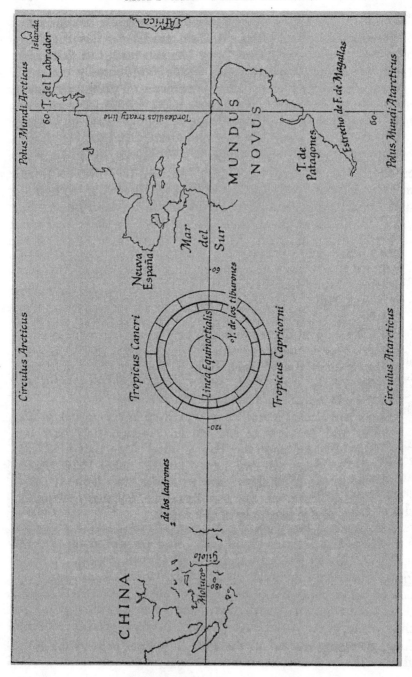

The Pacific Ocean redrawn from the chart by D. Ribero, 1529

incidentally, closely resembles one of Rodrigues' charts. The distance between the Asian mainland and the Moluccas has been reduced, and the total result is to put them 172° 30′ W. of the Tordesillas demarcation line, that is, seven and a half degrees within the Spanish sphere. This being the result aimed at by the Spaniards, it possibly explains the retention of the eastern prolongation of Asia. In the west the width of the Atlantic along the tropic is very accurate, but the width of the Pacific is of course reduced to accord with the position assigned to the Moluccas by about 11°. It would be interesting if this chart could be compared with one supporting the Portuguese case; no such map, however, appears to have survived.

Other features of Ribero's map are the approximately correct length of the Mediterranean; the distortion of northeast Africa, probably due to accumulated errors arising from neglect of the magnetic declination, which left a grossly exaggerated distance between the Red Sea and the Mediterranean; and the representation of the eastern coasts of North and South America as continuous. The Rio de la Plata is shown in detail, with three main affluents. The conspicuous error here is the exaggerated extent in longitude of the northeast coast of South America, the perpetuation of an early mistake which persisted right through the seventeenth century. It is possible that it arose from this section having been charted originally on a larger scale than the adjoining Caribbean area. A small portion only of the western coastline is shown, based on the Balboa and Pizzaro exploits. The portions resulting from Magellan's voyage are the coastline south of the Rio de la Plata and the Straits of Magellan, the islands 'de los ladrones', rather curiously placed in 12° 30′ north latitude instead of 2° N., and an uncompleted group of islands representing the southern Philippines and the north coast of Borneo.

Ribero's placing of the Moluccas $7\frac{1}{2}$° within the Spanish sphere represents the last position taken up in the dispute by Spain, who had begun by claiming that the meridian ran through the Ganges delta. In the year the chart was made the Spanish crown, in view of all the uncertainties, sold its claim to the Portuguese—a good bargain since it was untenable.

What effect had all this activity of the seamen and chart

makers upon the cosmographers? As might be expected, they began by attempting to fit portions of the new discoveries into the conventional framework, and finished by accepting unreservedly the new pattern of the world revealed by the navigators. Three stages in this process may be discerned: the emendation of a world map which had much in common with that used by Martin Behaim for his globe; an intermediary stage which produced a combination of Ptolemaic and the 'new' geography; and finally the adoption of the complete contemporary world outline as embodied in the Canerio chart. This transformation was made, as far as printed maps are concerned, in the space of ten years, as can be seen in the maps of Martin Waldseemüller.

The first in this series is a map of the world, designed by Giovanni Matteo Contarini and engraved on copper by Francesco Roselli in 1506, a unique copy of which is in the British Museum. The map, on a conical projection with Ptolemy's prime meridian as the central meridian and the Equator truly drawn, has the eastern coasts of Asia in the west and Ptolemy's Magnus Sinus and the islands of the medieval travellers in the east. In one of the inscriptions the cartographer says: "if by folding together the two sets of degrees [i.e. on the eastern and western margins] you form them into a circle, you will perceive the whole spherical world combined into 360 degrees". This is not strictly true, for the map does not extend much beyond the Tropic of Capricorn; but elsewhere there are verses extolling Contarini for having marked out

"The world and all its seas on a flat map,
Europe, Libya, Asia, and the Antipodes,
The poles and zones and sites of places,
The parallels for the climes of the mighty globe."

These references to the whole sphere, the Antipodes, the poles, and the globe, are intriguing; it is possible that the cartographer, especially in view of some similarities between his map and Behaim's globe, had in fact a globe before him. It is possible, but not very probable, that another section of his map, now lost, portrayed the southern hemisphere. The map incorporates a good representation of Africa, with the names fairly close to those of Cantino, and an attempt to fit in the India of Vasco da Gama; between the Persian Gulf and the Indus of Ptolemy, the cartographer has inserted a narrow peninsula, trending south-westwards, on which are shown the towns of Cōbait (Cambay), Cananor and Calicut (these two were visited by Vasco da Gama). Correctly placed in reference to this Indian peninsula is the island Seilā (Ceylon). To the east, however, the Ptolemaic outline is retained, including the great island of Taprobana, which was originally also Ceylon. To increase the confusion, a 'Seilā īsula' also appears among the south-east Asian islands, probably standing for Sumatra. This confusion also occurs on the Behaim globe.

The western portion of the map is perhaps the most interesting, for the extent to which it illustrates the ideas of Columbus. The east Asian coast is similar to that of the Behaim globe; the north-east peninsula extends, however, to within twenty degrees longitude of Europe, and on its eastern extremity are represented discoveries attributed to the Portuguese (evidently Cortereal). Fifty degrees east of Asia, and on the Tropic of Cancer, appears Zimpangu, which is stated to be identical with Hispaniola. Between Zimpangu and the west African coast, the discoveries of Columbus and the Spaniards are inserted, the group of islands, Terra de Cuba, Insula Hespaniola, etc., with no suggestion of a North American continent, and the north-east coast of South America as discovered by Columbus on his third voyage and his Spanish successors. The representation here shows Spanish influences, and Heawood did not consider that the Cantino chart was a direct source. An interesting feature is that a conventional western coastline has been given to this southern land-mass. Perhaps this is intended to be the antipodean continent suggested in the verses quoted above.

Two years after the Contarini map, another very similar to it was published at Rome, and is found in copies of the 1508 Ptolemy edition. This is attributed to Johannes Ruysch. Except for small details, the projection is identical with that of Contarini's map. It is stated to be 'ex recentibus confecta

D

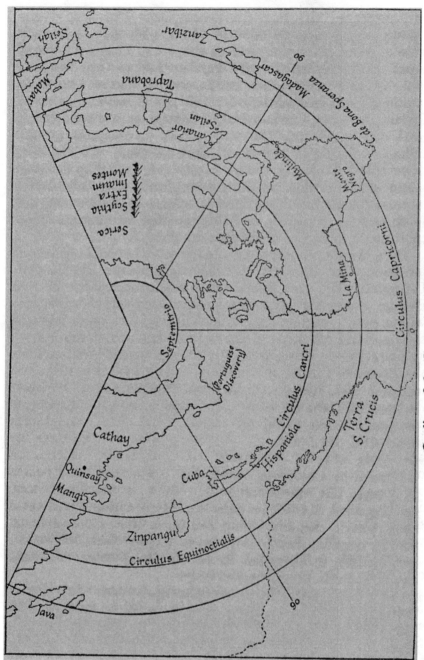

Outline of the Contarini World Map, 1506

observationibus', and certainly draws on sources later than Contarini. The Indian sub-continent has much better proportions, but the Far East is in general still Ptolemaic, and the three 'Ceylons' occur again. The inscription identifying Zimpangu with Hispaniola is repeated, but there is an interesting addition; 20° west of the Azores is inserted 'Antilia insula', the mythical island in the Atlantic, which first appears on charts of the early fifteenth century.

In South America there are also important additions. The eastern coast is continued southwards to the 'Rio de Cananor' in 30° S., and it is noted that exploration has extended to 50° S. latitude, a reflection of Amerigo Vespucci's voyage of 1501.[1] In the north, there is an isolated portion of the mainland, probably Florida, and the Portuguese discoveries in the far north, with the addition of Greenland, are again shown as part of Asia.

With the Ruysch map, a conventional representation of the world, current from the 'eighties of the previous century, disappeared from general circulation. Its place was taken in geographical circles by conceptions popularized by Martin Waldseemüller in his great world map of 1507, and his 'Carta marina' of 1516. The world map of 1507 is a massive woodcut in twelve sheets, on a single cordiform projection. Its title accurately describes it as "according to the tradition of Ptolemy and the voyages of Amerigo Vespucci and others". (As is well known, it was Waldseemüller who in his 'Introductio' proposed the name America for the newly found western lands.) Joseph Fischer and F. von Wieser showed conclusively in their memoir on the unique copy of this map, that the source for the new discoveries was the Canerio world chart, and not merely a copy of this but the actual surviving chart. The south-east coast of South America is carried to 50° S. (cf. the notes on the Ruysch map). The eastern coast of the Central American isthmus is inserted, separated by an extremely narrow strait from the small portion, extending a little north of Florida, of the northern mainland, which is also represented. Northern Africa and Asia

[1]For this voyage, and the cartography of South America in general, see the important study by Roberto Levillier, 'America la bien llamada', 2 vols. Buenos Aires, 1948.

are after Ptolemy but south-east Asia retains features of the
Contarini-Ruysch type. One thousand copies of this map were
printed, a large edition for the time, and evidence of the intense
interest of Europe in the new discoveries. Waldseemüller could
record with satisfaction that it was received with great esteem.

Owing to its essentially Ptolemaic basis, the map gives an
extremely exaggerated representation of the eastward extension
of Asia; in fact the land-mass of the Old World extends through
some 230 degrees of longitude. Soon after its publication,
however, Waldseemüller appears to have adopted the new
views of the navigators, for included in the Strasbourg edition
of Ptolemy, 1513, is a crudely drawn version of the Canerio
chart, with a few names only—the 'Orbis typus universalis
iuxta hydrographorum traditionem'. This but foreshadowed
the monumental 'Carta marina navigatoria Portugallen' of
1516 (the Spaniards and others are rather unkindly overlooked)
in twelve woodcut sheets, and in view of what has been said
already about the Canerio map, its context requires little
comment: as its author states, it contains features "differing
from the ancient tradition, and of which the authors of old were
unaware". The most striking feature is the reduction of the
longitudinal extent of Asia to something approaching reality.
Compared with the map of 1507, it exerted little influence on
later cartographers, though a poor second edition, with legends
in German, was published by Laurentius Fries in 1525. On the
other hand, the 1507 map became for three decades at least the
accepted world type; Schoner's terrestrial globe of 1515
follows it very closely, and in 1520 Peter Apian produced a
greatly reduced version, without acknowledgment, thus gaining
for himself an undeserved reputation. Versions of the latter
were edited by Gemma Phrysius and Sebastian Münster, so
that the Waldseemüller type held the field until the advent of
Mercator, Ortelius, and the Dutch school.

REFERENCES

CONTARINI, G. M., A map of the world designed by G. M. Contarini, engraved by F. Roselli, 1506. (Brit. Mus.) 1924. Quotations by permission of the Trustees, British Museum.

CORTESÃO, A. Z., Cartografia e cartógrafos portugueses dos séculos XV e XVI. 2v. Lisboa, 1935.

—— Hitherto unrecognized map by Pedro Reinel in the British Museum. (*Geogr. Journ.*, 87, 1936, 518.)

——, Cartografia portuguesa antiga. Lisboa, 1960.

FONTOURA DA COSTA, A., A marinharia dos descobrimentos, Lisboa, 1933.

HAMY, E. T., La mappemonde de Diego Ribero, 1529. (*Bull. de géogr. hist.*, Paris, 1887.)

HARRISSE, H., The discovery of North America, 1892.

HEAWOOD, E., The world map before and after Magellan's voyage. (*Geogr. Journ.*, 57, 1921, 431.)

—— A hitherto unknown world map of A.D. 1506. (*Geogr. Journ.*, 62, 1923, 279.)

SKELTON, R. A., The cartography of Columbus' first voyage. (In Vigneras, L. A., Journal of Christopher Columbus, London, 1960.)

—— The cartographic record of the discovery of North America; some problems and paradoxes. (*Actas Congresso internac. dos descobrimentos* vol. 2, Lisboa, 1960.)

UHDEN, R., An unpublished Portuguese chart of the New World, 1519. (*Geogr. Journ.*, 91, 1938, 44–56.)

—— The oldest Portuguese original chart of the Indian Ocean, 1509. (*Imago Mundi*, 3, 1939, 7.)

WIESER, F. VON, Die Karte des Bartolomeo Colombo. Innsbruck, 1893.

CHAPTER VII

TOPOGRAPHICAL SURVEYS OF THE FIFTEENTH AND SIXTEENTH CENTURIES

As we have seen, new maps of various European countries began to appear in editions of Ptolemy before the end of the fifteenth century. These are in general based on outlines from the marine charts, with contemporary names in place of Ptolemy's, and some additional features. The latter were probably drawn from regional maps, which were being made in northern Italy as early as the fourteenth century.

Rivers are perhaps the easiest natural features to map fairly accurately without elaborate instruments. They were also a principal means of transport, and would be well known to generations of watermen. At a later date, maps would be useful when the need for controlling or improving their courses arose. From the fifteenth century quite elaborate maps of territories in northern Italy have survived. The territories of the city states, often compactly grouped around the capitals in well-watered plains, formed reasonably sized units for the delineators. One of the earliest of these extant 'surveys' is an exceptionally large one (3.05 m. × 2.25 m.) of Verona and its territories, ascribed to c. 1440. This is carefully drawn and coloured, with mountains in brown, rivers a green-blue, vegetation light green, roads yellow, and names in red. It seems obvious that the style of execution is the culmination of a long tradition, and not something evolved in a few years. As to the methods employed in drawing the map, Professor Almagià suggests very reasonably that it was built up on distances and directions radiating from Verona; along the main roads and near the middle of the map, the accuracy is fair, and distortions are greater near the margins, particularly in areas not traversed by roads. Another element in this map has been supplied by the topographical artist; along the lakes, for example, profile views of the mountain ranges have been drawn, and the

102

mountains generally are rendered realistically; Verona is represented by a bird's-eye view, in some detail, the smaller towns by a few buildings, drawn 'on their backs'. The part of the topographical draughtsman in determining the conventions of such maps was considerable and continued certainly for two, perhaps for three, centuries. Such surveys were becoming appreciated by administrators, and in 1460 the Council of Ten of the Republic of Venice ordered all commanders of cities, lands, and fortresses to send maps of their ,jurisdictions to Venice. Almagià suggests that this order was responsible for a surviving map of Padua, 1465, and one of Brescia, c. 1470. The former, perhaps by the painter Francesco Squarcione, is more stylized than the Verona map; towns are generally represented by single towers, and the canals, or *tagli*, are included. The map of Brescia is notable for the accurate and complete delineation of the relief and particularly the hydrography; roads and bridges also are well done; but again details in less accessible areas are sketchy. These are but a few of the many MS. local surveys which must have existed in fifteenth-century Italy, and undoubtedly provided material for the engraved maps which became so numerous in the following century.

It is possible that the art of cadastral survey never died out completely in Italy. At any rate the designs which the Roman *agrimensores* employed to represent, for example, mountains, seem to have persisted continuously, for similar forms are found in the early Ptolemy manuscripts. Practical requirements at a later date stimulated progress in the Netherlands where the earliest document which may be called a map is a sketch of part of the Oude Maas, dated 1357. In some cases these 'maps' were more nearly eye-sketches than pieces of survey, and in fact, before survey was established on a scientific basis, the landscape painter—or topographical draughtsman—played an important role in cartographic development. In the Netherlands the earliest 'maps' containing much detail were mostly oblique views drawn by landscape painters from church towers or other vantage points. Indeed in these early days it is difficult to distinguish between maps and 'bird's-eye views'. In charts of the coasts, this drawing of oblique views and

profiles of important landmarks persisted for many years.
Some of the earliest of the English coasts were drawn in this
fashion by Richard Popinjay about 1563.

The next stage was the adoption of methods based on
elementary geometry, the study of which was being developed
by the astronomers, using translations of Arabic texts. In the
second half of the fifteenth century, the University of Vienna
was an important centre of astronomical and mathematical progress. This was in great part due to the work of Georg Peurbach
(1423–61) and his pupil, Johannes Regiomontanus (1436–73).
These men were interested in geography through astronomy,
which led to the consideration of determining positions on the
earth's surface. Regiomontanus visited Ferara in the 1460s,
where he was captured by the current passion for Ptolemy's
'Geography', and projected a world map and new maps of
European countries. Later he translated the first book of the
'Geography' into Latin. His great work was done at Nurem-
berg in the last three years of his life, where he compiled a
calendar, his famous 'Ephemerides', or astronomical tables
much used by navigators, and a list of geographical positions,
largely derived from Ptolemy. He also compiled tables of sines
and tangents, in pursuing his aim to make trigonometry useful
to astronomers, and wrote the tract, 'De triangulis', dealing
with plane and spherical triangles, which introduced a new
era in the development of trigonometry.

A little later another celebrated astronomer and mathematician, Peter Apian, who spent five years as a student at Vienna
before he became a professor at Ingoldstadt, was associated
with the production of a number of maps, including one of the
world on a heart-shaped projection after Waldseemüller and
another of Europe, as well as regional maps. His main work
was in astronomy where he improved several instruments, and
advocated the determination of longitudes by lunar distances.
It is probable that men such as these, specialists in geometry,
trained in instrumental observation, and, to some degree, also
instrument makers, would have grasped the application of
simple geometrical operations to rudimentary survey. In 1503,
the encyclopaedic 'Margarita philosophica' of Gregor Reisch
contained a description of the 'geometrical square'—a square

with a graduated circle and a movable sighting arm (or alidade) attached. The text explains how with this instrument it is possible to determine the mutual relation of towns, i.e. by observing the bearings of one from the other. For this it would be necessary to orientate the instrument correctly at each point, and Waldseemüller explained, some years later, in the booklet accompanying his 'Carta itineraria', how this could be done with a 'compass clock', that is, a combination of a sun dial and compass. In the 1512 edition of the 'Margarita philosophica' the instrument has become more elaborate; called a 'polymetrum', it consists in essence of the geometrical square and alidade with a quadrant erected upon it, so that vertical and horizontal angles could be observed. Some have seen in this the prototype of the theodolite.

The famous cosmographer and cartographer, Sebastian Münster, while at Heidelberg University, became acquainted with the 'Margarita philosophica' and its rudimentary survey instructions. In 1528, he published an appeal to his fellow scholars to co-operate with him in a geographical description of Germany, which he proposed to supplement with an atlas.

"It is known and apparent that the regional maps of Germany, as they have been issued in recent years, are not constructed with correct observation of the azimuth, as is well seen with the great bend of the Rhine between Strasburg and Mainz which in truth is not set down as I have many times observed it."

He suggested that each of his friends should undertake to map the country within a radius of six to eight miles of their town, and he described how this could be done. With the instrument, in this case a quadrant divided in seventy-two sections and oriented by a compass, the observer took the bearing of a neighbouring village, drew a corresponding ray upon a sheet of paper and marked off on it to scale the distance between the two places. This operation was to be repeated at each village or observation post. Münster did not apparently contemplate fixing positions by intersecting rays or calculating distances from the triangles, without direct measurement. He

D*

concludes with confused directions for determining the latitude of the central town, and the book includes a small map of the environs of Heidelberg, given as an example.

The method of elementary triangulation is for the first time described more or less clearly by Gemma Frisius in his 'Libellus de locorum describendorum ratione', included in his edition of Peter Apian's 'Cosmographia', 1533. Gemma describes the operation in terms very similar to Münster's, and the instrument was probably much the same; he insists strongly on the necessity of placing the compass on the 'planimetrum' to orientate it correctly. But he goes further in fixing the position of places by intersecting rays, and in showing that the measurement of one side of a triangle will fix the scale of the map. He illustrates his theories with a diagram of an actual survey by this method between Brussels and Antwerp.

In the following years this method was practised extensively and refinements introduced. If it is referred to as 'triangulation', it should not be interpreted in the modern sense, of a base line measured with extreme accuracy and a system of well-conditioned triangles built up on it by careful angular measurements. The sixteenth-century usage was somewhat rough and ready, though it anticipated the use of the plane table for filling in detail. In combination with observations for latitude and calculation of longitude, it was capable of producing a map of considerable accuracy. Perhaps the best-known practitioner of the method was Philip Apian, son of the celebrated astronomer, who surveyed Bavaria between 1555 and 1561. Like all cartographers of his day, Apian was at pains to keep his methods secret from his rivals, but there is a good deal of contemporary evidence about them. The calculation of the lengths of sides of a triangle from one known side and the respective angles was clearly set out by Christoph Puehler in a book known to Apian, and one of his pupils, G. Golgemeier, describes in detail the procedure of mapping a small area. In his petition to Duke Albrecht for a privilege to publish his map, Apian complains of the expense to which he had been put. He had traversed Bavaria in the course of six summers, with companions to be supported and three horses to be maintained. Moreover, he had been obliged to summon the 'oldest inhabitants' of many places to Munich to obtain necessary details. Notes for a portion of his survey have survived, mainly lists of angles observed. We may therefore picture him, riding through the countryside in the summer months, with two companions, and periodically halting to ascend a convenient church tower or hill with his diopter, compass and notebook and to take angles to all prominent features in the surrounding countryside. From his notes it appears that, in an area of approximately 25 kms. × 35 kms. he observed at twenty-eight stations, taking in all 200 sights. Some of his rays were quite long, even by modern standards, up to fifty kilometres. The winter months were spent in working up his observations and drawing the map, and also in eliciting details from experienced countrymen.

The basis of Apian's map was his calculations of the latitude and longitude of a number of important towns, the latitudes being obtained from the observation of the passage of circumpolar stars. As for the longitudes, Peter Apian had advocated the use of lunar distances, but these observations were not accurate; his son therefore calculated longitudes from the differences in latitude and the direct distance between two places. These positions were entered in the map net and served as centres for the detail collected by Apian on his peregrinations.

It has been shown that he kept mainly to the valleys, and that the intervening plateau or mountain was sketchily represented, and often served to absorb the accumulated errors. He seems occasionally to have measured approximately the side of a triangle as a check on his observed angles; this was probably calculated from riding time between the two points, for he expressly states that one hour's riding equalled one German mile.

By these methods, he attained to a considerable degree of accuracy, at least within the area of his direct observation, and his work remained the basis of all maps of Bavaria for more than two centuries. His angles in general were accurately observed, but errors in his determination of latitudes affected the system of co-ordinates.

It will be noticed that Apian appears to have entered the

details of his observations in his notebooks, and then drawn out his map from them in the winter months, though others may have plotted the rays in the field. The former method had obvious drawbacks, and made it easy for errors to creep in. About this time, however, an advance was made which enabled the surveyor to draw his map as he went along. Instead of using the sighting rule mounted on the horizontal circle, divided in degrees, the surveyor placed the sighting rule directly on his drawing paper, mounted on a table, and aligning it on the distant object, ruled in his ray directly: provided he was careful to orient his table correctly at the successive stations, he could obtain as accurate results as with the horizontal circle. This method was referred to by Leonard Digges when he wrote his 'Pantometria', 1571:

"Instead of the horizontal circle, use only a plaine table or boarde whereon a large sheet of parchment or paper may be fastened. And thereupon in a faire day to strike out all the angles of position each as they find them in the field without making computation of the Grades and Scruples" (i.e. without observing directly the degrees and fractions).

This, as Digges' phrase in fact suggests, was the origin of the plane table, which was later greatly developed for filling in detail around the fixed points. When the technique was fully evolved, the plane table virtually became a survey instrument. Some time before 1570, William Bourne was using these methods around Gravesend and Tilbury.[1]

The well-known extract from the Privy Council warrant issued to Christopher Saxton when he was embarking on his surveys of the English and Welsh counties suggests that his methods had some resemblance to those of Apian: he was to be "conducted unto any towre, castle, high place or hill to view the country .... accompanied with two or three honest men such as do best know the country for the better accomplish-

[1]Taylor, E. G. R., William Bourne; a chapter in Tudor geography. (*Geogr. Journ.*, 72, 1928, 335.)

ment of that service". In Wales, incidentally, he was to be provided with "a horseman that can speke both Welsh and Englishe to safe conduct him to the next market town". No doubt the honest men would name the prominent features visible from the observation point for Saxton's information.

REFERENCES

ANDREAE, S. J. F. and VAN'T HOFF, B., Geschiedenis der Kartografie van Nederland. The Hague, 1947.

CRONE, G. R., Early maps of the British Isles, A.D. 1000–1579. (*Royal Geogr. Soc.*, London, 1961.)

GASSER, M., Zur Technik der Apianische Karte von Bayern. (*Verhandl.* 16. *Deut. Geographentages, Nürnburg*, 1907, 102–23.)

HANTZSCH, V., Sebastian Münster. Leipzig, 1898.

LYNAM, E., An atlas of England and Wales. The maps of Christopher Saxton, engraved 1574–1579. Introduction. (*British Museum*, 1939.)

POGO, A., Gemma Frisius . . . and his treatise on triangulation. (*Isis*, vol. 22, no. 64, 1935, 469–85.)

TAYLOR, E. G. R., The plane-table in the sixteenth century. (*Scottish Geogr. Mag.*, 45, 1929, 205.)

WEISZ, L., Der Schweiz auf alten Karten. Zürich, 1945.

CHAPTER VIII

MERCATOR, ORTELIUS AND THEIR SUCCESSORS

WITH the progress of exploration and the growing demand for topographical maps by travellers, statesmen, merchants and antiquarians, numbers of maps, large and small, flowed from the presses as the sixteenth century advanced. To the teachers of cosmography at the universities—or failing them, the publishers and their assistants—largely fell the task of co-ordinating and generalizing this varied material. This could be done by revising the world maps of earlier decades, which often appeared in many sheets, and were susceptible to damage or destruction, as the few copies of them which have survived testify. The varying sizes, too, of the smaller maps of continents, countries, provinces and counties made them awkward to preserve conveniently in bound volumes.

In the early years of the century, the only approach to a modern atlas as known today was the Waldseemüller edition of Ptolemy, with its twenty 'tabule novae'. The 'Cosmographia' of Sebastian Münster, Basel, 1550, contained what may be regarded as an atlas supplement of rather crude woodcut maps, some deriving ultimately from Waldseemüller, but others of special regions supplied by his friends. In Italy it had become the practice to bind up some of the finely engraved maps published at Venice and Rome to suit the tastes of individual collectors. The map engraver and publisher, Antoine Lafreri, established at Rome, had issued an engraved general title page for such volumes—'Geografia: tavole moderne di geografia de la maggior parte del mondo', 1560–70. These so-called Lafreri atlases sometimes included reduced copies of large maps, which are otherwise unknown or extremely rare. An important example is Olaus Magnus' 'Carta marina' of 1539, a map of the countries of northern Europe which was republished on a smaller scale by Lafreri in 1572. But it was

110

the Flemish cartographers Ortelius and Mercator who, in addition to their other achievements, met in a practical way the public demand for a comprehensive, up-to-date, and convenient collection of maps, by inaugurating the long series of modern atlases.

Gerhard Mercator (the Latinized form of his surname, Kremer), born at Rupelmonde in Flanders in 1512, owed much to his relations with Gemma Phrysius, the cosmographer and editor of Peter Apian. As a pupil of Gemma's at the University of Louvain, Mercator showed himself to have an aptitude for practical tasks. He is first mentioned as the engraver for the gores of Gemma's globe of about 1536; he was also a maker of mathematical and astronomical instruments, and in his early days a land surveyor. It was no doubt this aptitude that led him later to examine and to solve the problem of concern to the practical navigator, namely the representation of constant bearings (loxodromes) as straight lines on a chart. In the course of his long life, he also acquired a profound knowledge of cosmography and of topographical progress in Europe and beyond, and won general recognition as the most learned geographer of his day. While at Louvain, he established himself as an authority on all these matters in the intimate circle of the Emperor Charles V, a position which, in particular, brought him into contact with the navigators and cartographers of Portugal and Spain, then in the van of progress in these sciences. His principal achievements were his globe of 1541 and his celebrated world map of 1569; his large map of Europe of 1554; his edition of Ptolemy, 1578, and his Atlas, still in course of publication at his death in 1594.

The practical seaman of the day required a chart on which a line of constant bearing could be laid down as a straight line. This was impossible on contemporary charts, which made no allowance for the convergence of the meridians. If a line is to preserve a constant bearing on the globe it must cut each meridian at the given angle. Since the meridians converge on the Pole, this line clearly becomes a spiral, circling closer and closer to the Pole, but theoretically never reaching it. On his globe of 1541, on which for the first time these loxodromes were laid down, Mercator put them in by means of a simple drawing instrument which could be set at the required angle. But the problem of representing these as straight lines on a flat chart remained to be solved. The claim that Mercator was the first to recognize the true character of loxodromes has been disputed. The celebrated Portuguese mathematician and expert in the science of navigation, Pedro Nunes, was already investigating them, and in view of the close relations between Portugal and Flanders at the time, it is quite probable that Mercator was aware of his work. As far as is known, however, Nunes never reached the stage of projecting a chart on which they could be laid down as straight lines. This was finally done by Mercator through his great world map of 1569, on the projection now known by his name. Nunes, it may be noted, was highly critical of the charts of his time, complaining for instance that the pilots persisted in attempting to express distances in degrees instead of setting down the actual run in leagues, thus introducing endless confusion.

Before the appearance of his famous world map in 1569, Mercator had achieved an international reputation as a cartographer, principally through his map of Europe of 1554, which displayed critical ability of a high order. This map, of which only one copy is now known, was engraved in fifteen sheets, with over-all dimensions of 132×159 cms. It was published at Duisburg, where Mercator had established himself as a map maker and lecturer at the university in 1552. The map is a fine piece of engraving, with the lettering in the italic style which he popularized in Western Europe.

The principal improvement he effected was in the reduction of the length of the Mediterranean. Ptolemy's figure of approximately 62° had been generally followed by cosmographers. Mercator accepted Ptolemy's position for Alexandria, but established from the marine charts that the Canary Islands, through which the Alexandrian's prime meridian ran, were much further west of the Straits of Gibraltar than had previously been recognized. Consequently by allowing for this and by revising other distances, he reduced the longitudinal length to approximately 52°. This, though still about 10° 30′ in excess of reality, was a considerable advance. For over a century and a half, this was not improved upon by map makers, though the navigators had a more correct idea. In contrast to the longitudes, the latitudes on the map are quite accurate for western Europe, but towards the north and east errors of 2 to 3 degrees occur. This calculation is typical of the kind of reasoning upon which Mercator had to construct his map. Having accepted Ptolemy's position for Alexandria, he arrived at the positions of the principal points by careful research into distances from the best itineraries procurable, paying also particular attention to relative directions, in which he was helped considerably by the marine charts. The results thus obtained he co-ordinated to the best of his ability with the known latitudes of the principal cities. In a note on the map he dismissed the attempts to calculate differences of longitudes by the simultaneous observations of eclipses, for the perfectly sound reason that the precise moment of eclipse is extremely difficult to observe. An error of four minutes in determining this would throw the result out by one degree of longitude. Another of his improvements rendered the 'waist' of eastern Europe, between the Baltic and the Black Sea, much more accurately; on earlier maps, it had been far too constricted. On the other hand, the contour of the Black Sea is elongated by several degrees.

These brief comments will show the general methods of compiling maps of larger areas in the sixteenth and seventeenth centuries, and will suggest the kind of errors that might occur. Maps depended largely upon the labour expended in the cartographer's office in attempting to reconcile a mass of disparate and often conflicting data. Outside Europe the only features in the maps of the continents that were at all reliable were the coastlines, obtained from the marine charts. A partial exception was Asia, though even there knowledge of the interior parts was often antiquated and confused. This continued for long to be the general position, until in the nineteenth century explorers and travellers were equipped with reasonably accurate instruments for the rapid determination of positions, and the work of precise survey within modern limits of accuracy was gradually extended. Even at the present time, much of the earth's surface is still unmapped to this standard. This is one consideration which should be constantly borne in mind in discussing the work of map makers for at least two centuries after Mercator.

The unique quality of his map of Europe was at once recognized, and for the period the demand was large. A second edition was published in 1572, with considerable improvements especially in the northern regions. There Mercator was able to use the results of the English voyages to the White Sea, and English observations for the latitude of Moscow, combined with itineraries for the interior of Russia. Another important work dating from this period was his map of the British Isles of 1564. Oriented with the west at the top, it measures 129 × 89 cms. The compiler is unfortunately unknown, Mercator merely stating that he engraved it for an English friend.

Mercator's posthumous fame rests upon his world map published at Duisburg in 1569: 'Nova et aucta orbis terrae descriptio ad usum navigatium emendate accomodata'. This great map, of which only four copies have survived, is made up of twenty-four sheets in all, its full dimensions being 131 × 208 cms. Though the title refers only to its use for navigators, Mercator states that it was also intended to represent the land surfaces as accurately as possible, and to show how much of the earth's surface was known to the ancients.

As has been seen above, lines of constant bearing on the surface of the globe must be spirals, which ultimately circle round the Pole. To represent them as straight lines on a flat map, the meridians and parallels must be arranged so that loxodromes cut the meridians at constant angles, i.e. the meridians must be parallel. Since the meridians in fact converge, the effect of this is to distort east-west distances, and therefore direction and area at any given point. If however the distances between parallels are increased proportionately to the increase in the intervals between the meridians from the Equator towards the Poles, the correct relationships of angles, i.e. direction, are preserved. This was the solution obtained by Mercator, and charts on his projection were thus said to have 'waxing latitudes'. The projection has a further useful property: since at any point, the angles are correct, the shape of small areas is preserved (i.e. the projection is conformal). This property combined with that of representing loxodromes as straight lines, makes it extremely useful for representing small areas.

For large sections of the globe, it has obvious inconveniences. Owing to the 'waxing latitudes', the scale increases progressively from Equator to Pole, and to measure distances is not a simple operation. (The length of a degree of longitude is zero at the poles, whereas on Mercator's projection it is theoretically the same as at the Equator.) Mercator therefore placed on his map two lengthy notes, explaining how, given two of the following elements, difference of latitudes, difference of longitudes, direction, and distance, it was possible to determine the other two. The main difficulty was in the determination of distance owing to the variation in scale. This he solved by the principle of similar triangles. The triangle given by the bearing between the two points on the chart, whose distance was to be determined, and their difference in latitude was constructed proportionally on the Equator. The length of the required line was then measured off in equatorial degrees, and the result obtained by multiplying the figure thus obtained by the appropriate number of miles, one degree being taken to equal fifteen German miles, sixty Italian, or twenty French miles.

It was many years before Mercator charts were generally adopted by seamen, who preferred rule of thumb methods. There was some complaint that on the original world chart the coastlines were not clearly shown, but it is difficult to believe that this in itself led to its early neglect. The theoretical construction of the projection was not clearly set out until Edward Wright published his 'Certaine errors in navigation' in 1599. Until charts of relatively small areas were constructed on its principle, its merits would not be recognized by navigators. By the end of the century, such charts were beginning to be drawn, but it was not until nearly a century after its invention that Sir Robert Dudley produced a collection of charts, all of which were on this projection, in his 'Arcano del Mare', 1646. At the end of the century the famous navigator, John Narbrough, could still write:

"I could wish all Seamen would give over sailing by the false plain charts, and sail by *Mercator's* chart, which is according to the truth of Navigation; But it is an hard matter to convince any of the old Navigators, from their Method of sailing by the Plain Chart; shew most of them the Globe, yet they will talk in their wonted Road."

Mercator was also interested in the problem of terrestrial magnetism, and accepted the general observation of navigators that the line of no magnetic variation passed through the Cape Verde Islands. Accordingly, "since it is necessary that longitudes of places should, for good reasons, have as origin the meridian which is common to the magnet and the world . . . I have drawn the prime meridian through the said Islands". As he was also aware that the magnetic variation differed from place to place, he concluded that there must be a magnetic pole "towards which magnets turn in all parts of the world", and he marked the position of this pole in the region of the modern Bering Strait.

In his outlines of the continents Mercator broke away completely from the conceptions of Ptolemy, though the latter's influence in the interior of the old world can still be traced. He recognizes three great land-masses, the old world (Eurasia and Africa), the New Indies (North and South America), and a great southern continent, 'Continens australis'. This was based on an idea, which derived ultimately from the Greeks, that in the southern hemisphere there existed another continent which was the counterpart of the 'inhabited world'. Evidence in support of this theory was also based on a misreading of Varthema and Marco Polo, from which it was concluded that continental land, the hypothetical regions of Beach and Lucach, lay to the south of Java Major. The portions of Tierra del Fuego seen by Magellan were incorporated into this southern continent, the coastline of which was brought northwards to the vicinity of New Guinea. Here it is not beyond the bounds of possibility that the map preserves some traces of early knowledge of the Australian coasts. South-eastern Asia is based fairly closely on the Portuguese discoveries. Most of the interior, however, is derived from Marco Polo's narrative, and the

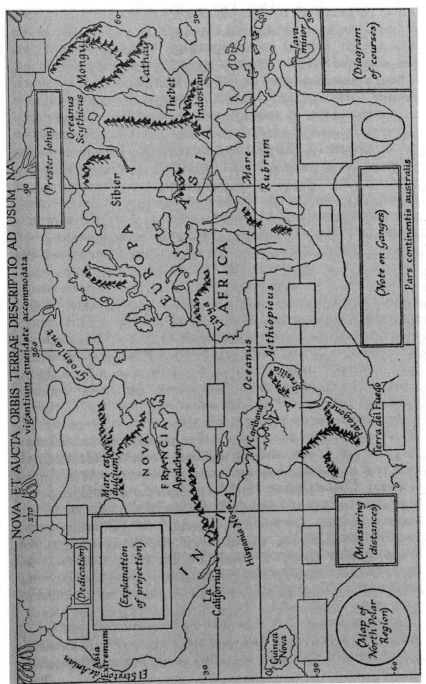

Outline of Mercator's World Map of 1569

outline recalls maps of the previous century and in part even the world maps of the later Middle Ages. The geography of the south-eastern interior is confused by Mercator's mistaken belief that the 'river of Canton' must be the classical Ganges.

South America has a curious quadrilateral outline, which was not corrected until Drake's voyage along the western coast. The width of the northern continent is considerably exaggerated; on the parallel of Newfoundland it amounts to 140° of longitude. On the west coast, California is correctly delineated as a peninsula; in the extreme north-west is the narrow strait, 'Streto de Anian', separating America and Asia, which has given rise to much controversy. In the interior, partly obscured by a cartouche, an area of water with the inscription "Mare est dulcium", indicates some knowledge of the Great Lakes, though too far north in relation to the St. Lawrence. The representation of the Arctic region is of particular interest. Mercator provided a special inset map, "as our chart cannot be extended as far as the pole, for the degrees of latitude would finally attain infinity". This shows open water at the Pole, surrounded by a roughly circular, independent land-mass. This representation was in part based on a report ('Inventio fortunatae') now lost, of a certain English minorite of Oxford, Nicholas of Lynn, who in 1360 travelled in these parts "with an astrolabe".

Mercator's map would provide the material for an almost endless commentary, but it must suffice to point out its influence on exploration. It was through channels round the Arctic land that it was hoped to find north-west and north-east passages to Cathay. Drake's plans for his voyage of circumnavigation envisaged the discovery and annexation of part of the southern continent, and his 'Nova Albion' lay in the 'Quivira regnum' of north-west America, conveniently placed to control the Strait of Anian. Tasman's circumnavigation of Australia was planned to determine its relation to the hypothetical continent, which continued to intrigue geographers until Cook reduced it to truer proportions.

This world map was regarded by Mercator as part of a co-ordinated scheme of cartographical research. The world map was to form the basis, and other sections were planned to deal with modern maps, the maps accompanying Ptolemy's 'Geography', and finally maps of ancient geography. The first to appear was his edition of the Ptolemy maps, 1578, mainly based on earlier printed editions. The maps were redrawn on a trapezoidal projection instead of the usual rectilinear projection, with a central meridian and two parallels at right angles truly divided. In the year in which the Ptolemy edition appeared, Mercator was already at work on the modern maps, but the labour on them was prolonged beyond his expectations, partly through difficulties in obtaining original maps and travel narratives and in finding engravers, and partly because he was obliged to support himself by other work. It was not therefore until 1585, when he was seventy-three years of age, that the first part of the collection, to which he later gave the title 'Atlas', appeared at Duisburg. This contained three sections, each with a separate title-page, covering France (Gallia), Belgium (Belgia Inferior) and Germany—fifty-one maps in all. Four years later, he published the second part—Italy, Slavonia, and Greece in twenty-two maps. Finally in 1595, a year after his death, his heirs published the complete work with a general title-page, 'Atlas sive cosmographicae meditationes de fabrica mundi et fabricati figura'. This was the first time the term 'Atlas' was applied to a collection of maps. The title-page shows the figure of a man bearing the globe upon his shoulders—though Keuning states that the name was derived from a mythical astronomer-king of Libya, who was said to have made the first celestial globe. The Latin sub-title of the third section, added to the two already published to form the complete work, may be translated "The new geography of the whole world". The section comprised thirty-four maps. Five of these were by Mercator's son, Rumold, and two grandsons; these consisted of maps of the world and of the four continents based on earlier works of Mercator. The remaining twenty-nine had been completed by him before his death; sixteen were of the British Isles, the remainder of northern Europe.

Mercator's atlas was not at first in great demand; one reason no doubt was the manner of its publication in sections which were really separate small atlases of individual countries.

Then when the whole appeared it was by no means complete, for it lacked maps of the Italian peninsula and individual maps of the world outside Europe. Later, when Jodocus Hondius bought the plates from Mercator's heirs, after a second, unaltered, edition had appeared in 1602, and added thirty-six new maps to supply some of the deficiencies, the Atlas, constantly enlarged, was very popular in the seventeenth century. After the first Mercator-Hondius edition, Amsterdam, 1606, some thirty editions appeared before 1640, a number with little alteration. In addition to the original Latin there were editions in French, German, Dutch, and English. As time went on, however, they became much more the work of the Hondius and Jansson families. Finally it was superseded by the atlas of William Janszoon Blaeu, the first version of which appeared in 1630.

It is necessary to retrace our steps at this point. The major factor in the little success which attended the early issue of Mercator's Atlas was undoubtedly the existence of Abraham Ortelius' 'Theatrum orbis terrarum', which had appeared as early as 1570.

Abraham Ortelius, born at Antwerp in 1527, was a scholar and craftsman rather than a practical cartographer. He began work as an illuminator of maps, and later established himself as a map seller. His business seems to have flourished, for he was able to indulge his classical tastes by forming a large library and a collection of antiquities. His extensive travels in western Europe, including the British Isles, brought him a wide circle of friends and correspondents, which included John Dee, William Camden, Richard Hakluyt, and Humphrey Lhuyd. Through these connexions Ortelius obtained a good deal of his material. Lhuyd, shortly before his death in 1568, sent him two maps of England, one with ancient and modern names, and another described as 'tolerably accurate', as well as a map of Wales. Dee wrote asking about his map of Asia, and Hakluyt asked him to publish something on North America. This was of course the period when the English cosmographers, merchants, and navigators were keenly interested in the practicability of the north-west and north-east passages to the East Indies.

Ortelius may have begun work on his collection of maps as early as 1561; he had at least issued some separately before 1570. These included a map of the world in 1563 and a map of Asia, largely based on the work of the Italian cartographer, Jacopo Gastaldi. There is a tradition that Mercator delayed the publication of his Atlas so that his friend Ortelius might have the honour of publishing the first uniform collection of maps, but this is probably apocryphal, since Mercator was another fifteen years in completing the first section of his Atlas.

The distinguishing features of the 'Theatrum' were critical selection from the best maps available, to give a comprehensive cover of the world; uniform size and style of the maps, specially designed for this publication; the citing of the authorities employed for each map; and the subsequent issue of the 'Additamenta' to supplement and bring the collection up to date. Ortelius' list of authorities, which contains the names of eighty-seven cartographers (ninety-one in the second edition), is an invaluable source for the history of cartography, and has formed the basis of an admirable monograph by Leo Bagrow.

The 'Theatrum', in its first form, contained seventy maps on fifty-three sheets, many engraved by Francis Hogenberg. The contents were: a map of the world, four maps of the continents, fifty-six maps of Europe (countries, regions, and islands), six of Asia, and three of Africa.

The 'Theatrum' was an immediate success, since it met, in a convenient form, the contemporary interest not only in the lands overseas but also in the topography, administrative limits, and antiquities of the nation states of Europe. A second edition appeared in the same year as the first, and in all forty-one editions were published, the last in 1612. In addition to the twenty-one Latin editions, there were two Dutch editions, five German, six French, four Spanish, two Italian, and one English (1606). At the death of Ortelius in 1598, the 'Additamenta' had added about one hundred maps, some with insets, and a number of the plates had been re-engraved or altered. Besides this, editions from 1579 onwards contained the 'Parergon', a series of historical maps which eventually constituted a historical atlas, the work of Ortelius himself.

Those from whom Ortelius drew his material represent the leading cartographers of his day, and their number gives some idea of the intense activity then displayed in this field. Some of the maps have already been, or will be, mentioned in this outline: among the others may be noted the maps of Westphalia and Gelders by Christopher Schrot; Flanders by Mercator; the Duchy of Austria, Hungary, Tirol and Carinthia by Wolfgang Lazius; Italy and Italian districts by Jacopo Gastaldi; Bavaria by Philip Apian; Switzerland by Aegidius Tschudi; Russia and Tartary by Anthony Jenkinson (first published in London in 1562) and Humphrey Lhuyd's maps of England and Wales. The originals of many of these maps were produced by methods similar to those applied by Philip Apian.

Mercator states specifically that in the compilation of his world map of 1569, he had used Spanish and Portuguese charts, and for some decades afterwards these remained the sole cartographic source for much of the New World and of the East Indies. When the Dutch, on breaking with Spain, began their overseas expansion, they were naturally at pains to obtain the best of these charts for the use of their pilots.

In search of sailing directions and charts, J. H. van Linschoten spent five years (1583–8) at Goa, and published the fruits of this visit in his 'Itinerario', Amsterdam, 1596. This was accompanied by maps of the East Indies based on the work of Luiz Texeira. On a similar quest, the brothers Cornelius and Frederick Houtman were sent to Lisbon in 1592. Portugal was then under Spanish control, and the brothers were thrown into prison, but eventually returned with twenty-five nautical charts obtained from the Portuguese cartographer Bartolomeo Laso. In making such charts available to Dutch navigators, an important part was played by Petrus Plancius, who had been involved in the Houtman episode, and who contributed a world map to Linschoten's 'Itinerario'. Plancius, a theologian and minister of the Reformed Church, gradually established himself as an expert on the route to the Indies and on navigation. For some time he was an advocate of the north-east passage, and was consulted on the preparations for Barentsz's voyage of 1595. His views were based largely on a map by Pedro de

Lemos, probably drawn in 1586 or later, which discarded the Mercator conception of the Arctic regions and showed a practicable northern route to the Indies. After the failure of these northern attempts, he concentrated on the African route, and prepared sailing instructions for the second Dutch East Indian voyage of 1598. Four years later he was appointed official cartographer to the Dutch East India Company. He published numerous charts, some of which have not survived, including charts of the Mediterranean on Mercator's projection, and he attempted to solve the problem of the determination of longitude by observing the variation of the compass.

During the seventeenth century the procedure was established that pilots returning from the east should hand over to the official cartographer their charts with the additions and amendments resulting from their own observations. The cartographer was responsible for collating these, and for preparing revised charts for succeeding voyages. A considerable body of such manuscript charts have survived as a witness of Dutch hydrographic activity. The information they contained was slow to find its way on to published engraved charts, probably as a matter of policy. Among Plancius' successors as official cartographers were Hessel Gerritsz and the Blaeus, father and son. Gerritsz was in 1622 the draughtsman of a magnificent manuscript map of the Pacific Ocean, largely based on Spanish sources but also incorporating the tracks of the circumnavigators Le Maire and Schouten. Among the charts which he engraved and published was the 'Caert van't landt Eendracht', 1627, which portrayed the coast of western Australia discovered by the Dutch vessel, *Eendracht*. Willem Jansz Blaeu, who succeeded Gerritz in 1633, was a student of mathematics and astronomy, and the founder of a famous cartographical establishment in Amsterdam, from which issued maps, atlases, wall maps and globes. Willem's tenure of office was short, and in 1638 his son, Joan Wz. Blaeu, succeeded him, and with his experience in his father's office, made a notable contribution to establishing standard charts for the Dutch navigators. With him the great work of his family virtually came to an end, for shortly before his death in 1673, his printing house and all his engraved plates were totally destroyed by fire.

Such men, in addition to being draughtsmen and cartographers, were also engravers and publishers and therefore catered for the informed public as well as for pilots. Reference has already been made to their activities as compilers and publishers of atlases. To conclude this brief sketch of the golden age of Dutch cartography something must be said of another characteristic feature of their work, the production of large maps of the world, suitable for mounting as wall maps.

The lead in this was taken by Petrus Plancius, who established a type which in many respects persisted throughout the period. In 1592 at the outset of his career as a cartographer he published a world map in eighteen sheets, with the considerable overall dimensions of 146 × 214 cms. This was based on two principal sources, Mercator's world map of 1569 and a manuscript map by the Portuguese cartographer, Pedro de Lemos. Plancius, it is noteworthy, abandoned the projection devised by Mercator in favour of the simple cylindrical projection (*plate carré*) of Lemos' map. Mercator's projection, apparently on account of the distortions it introduced in the Polar regions, was not popular with the general public; the cylindrical projection, of course, owing to its parallel meridians introduced distortions of its own, but Plancius offset this to some extent by engraving on the map the length of a degree of longitude at each degree of latitude. Also in order to display the Polar regions more correctly, he put on the map two insets drawn on the equidistant zenithal projection centred respectively on the North and South Poles. He adopted the Portuguese conception of the Artic regions in place of Mercator's, and also abandoned Mercator's delineation of eastern Asia, which had attempted to reconcile later knowledge with Ptolemy's outline, adopting in its place the more realistic Portuguese representation. He retained, however, and indeed emphasized, Mercator's great southern continent, a theoretical conception supported by very few actual discoveries. He also made further improvements on Mercator, following, for instance, maps in Ortelius' 'Theatrum' for the delineation of interior China. The map is placed within an elaborately engraved border, and is covered with inscriptions, some of considerable length, and vignettes of ships, native peoples, and sea monsters.

Though one copy only has survived (at Valencia, in Spain), the map was very popular in its day. It is sometimes said, probably erroneously, that an edition was published in London. Thomas Blundeville, however, in his 'Exercises', 1594, gave a full translation of all the legends, and confirmed it as the work of Plancius (for the map does not bear his name). In 1604 the map was entirely re-engraved with modifications by J. van den Ende. The main changes were the insertion of Barendtsz's discoveries in the Novaya Zemlya area, a more detailed representation of Guiana based largely on Sir Walter Raleigh's voyage, the insertion of Davis Strait, and amendments to the coastline of southern Africa and South America, based on Portuguese and Spanish sources. Inset maps of important straits and passages were also added.

It is remarkable that it should have been thought worth while to re-engrave completely so large a map, but there was evidently a large contemporary demand for such works. The following year, W. J. Blaeu published a large general map in two hemispheres on the stereographic projection, and Plancius a very similar map in 1607. A year later, Jodocus Hondius broke new ground by publishing one drawn on Mercator's projection. Perhaps because this projection was not popular, not being understood by the general public, he reverted in his next world map of 1611 to the two hemispheres on the stereographic projection. The series was finally closed with J. W. Blaeu's large world map on a similar design, issued to mark the Peace of Westphalia which closed the long war with Spain in 1648. Without going into great detail, certain characteristics of these maps may be indicated. In general, they make use of the Portuguese-Spanish outline made familiar by Plancius, to which the results of Dutch exploration and charting were progressively added. Little attention was paid to inland areas, even to those to which the scanty information available applied. It thus came about that while later examples had improved outlines, the interior might be less accurately represented than

on their predecessors. When allowance is made for this, Blaeu's
world map of 1648 may be regarded as marking the highest
achievement of Dutch cartography. Upon it are drawn the
coastlines of northern and western Australia, of southern
Tasmania, and of parts of New Zealand, representing the
achievements of Abel Tasman. The coasts of China are greatly
improved, no doubt from Dutch charts, and something like
the true form of the continental coast north of Japan begins
to emerge, as the result of the voyage of Maerten Gerritsz
Vries. In the Arctic Spitsbergen is partially represented, and
the work of the English in and around Baffin and Hudson
Bays is delineated, though the orientation of Baffin Bay is at
fault.

To Blaeu's credit must also be set the abandonment of the
great hypothetical southern continent and of the four great
islands shown by Mercator as surrounding the North Pole.
Against these must be set blemishes such as the faulty orientation of the Amazon and the Rio de la Plata, and the retrograde
step of delineating California as an island. A common error in
this type of map was the undue longitudinal extent allotted to
the continents, especially to Asia. This was partly due to
the authority which Ptolemy still exerted, and also to the almost
total lack of reliable observations for longitude. On this map
of Blaeu's, the extreme longitudinal extent of Africa is represented as approximately 80°, an exaggeration of some 12°;
Asia, eastwards of the Mediterranean, is allotted 90°, instead
of 85°, and South America 55°, instead of 46°. On the Hondius
map of 1608, these exaggerations are even greater (Africa by
nearly 16° and South America by as much as 27°); but Asia,
by some chance, is almost correctly represented. It is noteworthy that the elder Blaeu, shortly before his death, was
aware that the conventional length assigned to the Mediterranean was greatly exaggerated. Despite these defects the
map gives a recognizable outline to the continents, and little
further progress could be expected without a fundamental
advance in method, particularly in the determination of
longitude. There is another noteworthy aspect of these maps,
their high technical and artistic quality, which may well have
been their chief merit in contemporary estimation. To excel

their rivals in this respect was without doubt a principal aim of their publishers and engravers. The surfaces are filled with carefully drawn and executed compass roses, appropriate types of ships, scenes of native life, navigational instruments, and similar conceptions which appealed to the imagination of their designers. But it is particularly in the lettering that they display the engraver's art at its finest. The italic lettering employed by Hondius has never been excelled for its purpose. These maps, by reason of their content were virtually encyclopædias of contemporary geography, and fittingly crown the century of Dutch supremacy in cartography.

REFERENCES

AVERDUNK, H. and MULLER-REINHARD, J., Gerhard Mercator und die Geographen unter seinen Nachkomen. (*Petermanns G. Mitt. Ergänzungsheft*, 182, 1914.)

BAGROW, L., A. Ortelii catalogus cartographorum. (*Petermanns G. Mitt. Ergänzungsheft*, 199, 210, 1928–30.)

DESTOMBES, M., La mappemonde de Petrus Plancius, 1604. (With facsimile.) Hanoi, 1944.

—— Cartes hollandaises. La cartographie de la Compagnie des Indes Orientales, 1593–1743. Paris, 1947.

DURME, M. VAN, Correspondance Mercatorienne. Antwerp, 1959.

HEAWOOD, E., The map of the world on Mercator's projection by Jodocus Hondius, 1608. (With facsimile.) 1927.

HEYER, A., Drei Mercator-Karten in der Breslauer Stadtbibliothek. (*Zeits. Wissenschaftl. Geogr.*, 7, 1890; 379.)

KEUNING, J., Petrus Plancius; theolog en geograaf, 1552–1622. Amsterdam, 1946.

—— The history of an atlas: Mercator-Hondius. (*Imago Mundi*, 4, 1948, 37.)

——, XVIth century cartography in the Netherlands. (*Imago Mundi*, 9, 1952, 35.)

WAGNER, H., G. Mercator und die ersten Loxodromen auf Karten. (*Ann. Hydrogr. u. Marit. Meteorol.*, 1915, 299.)

WIEDER, F. C., Monumenta cartographica. Vol. 2. Petrus Plancius, Planisphere, 1592. Vol. 3. Johannes Blaeu, World map, 1648. The Hague, 1926–8.

CHAPTER IX

THE REFORMATION OF CARTOGRAPHY IN FRANCE

In the seventeenth century, the desire to test new hypotheses of the physical universe stimulated attempts to determine accurately the dimensions and figure of the earth, and these then became possible through the invention of more precise instruments to make the necessary observations. These included the telescope, the pendulum clock, and tables of logarithms. The measurement of an arc on the earth's surface was the first step, and this, though primarily a geodetic operation, contributed eventually to the advance of cartography.

The first attempt of any value to determine the length of a degree in this way had been made by Snellius in Holland in 1615, but the operation was first carried out with an approach to accuracy in France, where in the latter half of the seventeenth century there was notable scientific activity under the patronage of the 'Roi soleil', Louis XIV, and the 'Académie royale des sciences' founded in 1666. In that country, one of the first highly centralized national states in Europe, there was a growing demand for maps and charts, and a realization that they could only be based satisfactorily on a precise scientific framework. Maps were required not only for military purposes, but for the proper organization of the extensive road system, the development of internal resources, which was the aim of men like Colbert, and the general promotion of commerce at home and abroad.

The successive stages in the making of the new map of France were (1) the measurement of an arc of the meridian of Paris by the Abbé Picard, 1669–70, by means of a chain of triangles; (2) the extension of the meridian until in 1718 it ran from the Pyrenees to Dunkirk; (3) the first attempts to produce a new map of France by adjusting existing surveys, supplemented by observations for latitude and longitude, to the Paris meridian; (4) the planned survey of the whole country, *de novo*, based on a complete system of triangulation, which resulted in the celebrated Cassini Survey. This had been advocated by Picard as early as 1681.

A remarkable contribution to this work was made by the four generations of the Cassini family. The first, Jean Dominique, who was invited to work at the Paris observatory in 1669, assisted in the determination of the meridian, but his greatest service to cartography was his perfection of a method of determining longitude by observation of the movements of Jupiter's satellites, a great improvement on the method of lunar eclipses, though the probable error was one kilometre. After the early work on the meridian, it was resolved to apply the new methods to rectifying the map of France, and Picard with other sur-

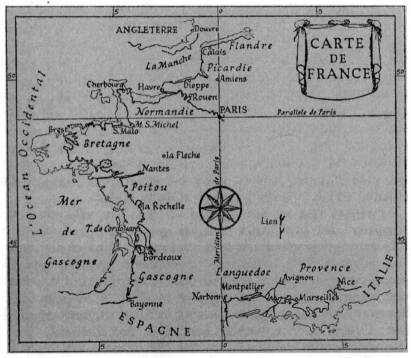

France, from the 'Carte de France corrigée ... sur les observations de MM. de l'Académie des Sciences', 1693, showing the revised position of the coast-line

E

veyors, including La Hire, known also for his projection, were sent to survey the coasts. A map by La Hire, embodying the results, was presented by him to the Académie in 1684, and subsequently published as 'Carte de France corigée par Ordre du Roi sur les Observāons de Mrs. de l'Académie des Sciences', 1693. This showed both the older outline of the coasts and the new, the general result being to shift the western extremity of France one and a half degrees of longitude to the east, in relation to the Paris meridian, and the southern coast-line about half a degree of latitude to the north. The sight of this map is said to have prompted Louis XIV to remark that the survey had cost him more territory than a disastrous campaign.

The second Cassini, Jacques, realizing that any attempts to fit haphazard surveys to the Paris meridian must be unsatisfactory, became the advocate of the complete triangulation of France, and with his son, César François Cassini de Thury, was engaged from 1733 on this extension. The backbone of the triangulation was the 'verified' meridian of Paris. Along this at intervals of 60,000 *toises* (rather more than one degree of latitude), perpendiculars were carried geometrically to the east and west, from which the positions of towns and other points of importance were fixed. This was the origin of the projection now known by Cassini's name, in which the co-ordinates of a point are given with reference to a central meridian and the distance along the great circle through the position which cuts the meridian at right angles.

Progress was set out in 1744 on the 'Nouvelle carte qui comprend les principaux triangles qui servent de fondement à la description géometrique de la France'. Cassini de Thury had succeeded in obtaining government support for the proposed topographical map on this framework, and the work was begun at the national expense in 1747. Nine years later, however, this was withdrawn, owing to heavy military expenditure. Cassini at once, without hesitation, took the bold step of assuming entire responsibility for the survey. He received authority to form an association to finance its completion, obtained the necessary support, partly from various provincial States-General who appreciated the value of accurate maps of their provinces, and carried the undertaking almost to completion. At the time of his death in 1784, Brittany alone remained to be published. Ultimately after suspension during the Revolutionary period, the work was taken over by the State and completed in 1818. Full details of the undertaking were given by Cassini in his 'Description géometrique de la France', published in 1783.

Improvements in instruments had greatly contributed to an improved standard of mapping. The divided horizontal semi-circles in brass were fitted with telescopic alidades, and micrometer reading allowed angles to be observed with considerable accuracy. Beacons, and sometimes lights, were used for observation marks. The topographic detail was treated more summarily: though the plane table was commonly in use by the 'ingenieurs géographes', the body of military surveyors, Cassini's men who carried out the minor triangulation sketched the details by estimation or by pacing, and worked this up in the office. Often they were content to indicate slopes by the letters D or F ('douce' or 'forte').

The Cassini map when complete comprised 182 sheets (88×55.5 cms.). The scale was $1:86,400$ (i.e. 1 inch to 1.36 miles). In style it is based on a map of the Paris region made in 1678, during the early days of the determination of the meridian, by Du Vivier, and engraved by F. de la Pointe. It is carefully engraved, the general effect being clean and uncrowded; the great 'routes' to Paris are emphasized and named, the larger towns are shown in plan, and a variety of symbols mark smaller settlements, churches, wind and water mills, gallows and other works of man. Forests, with their walks carefully drawn, are conspicuous, as are the residences of nobility and gentry, with their owners' names. Only in portraying relief does the map fail notably. In areas of lesser elevation, rivers and streams are depicted as flowing in narrow valleys with the borders hatched, and isolated elevations are only occasionally shown; the general effect therefore is of a vast level plateau dissected by canyon-like valleys. In the more rugged south and south-east, the result is even less satisfactory; the terrain is depicted in two or more tiers with the usual shading, and long crest lines show up as narrow white bands. The land-

forms of any considerable area thus appear curiously un-
integrated. It must be remembered, however, that it was not
for many years that sufficient determinations of altitude existed
for the relief to be represented with any accuracy.

Whatever the defects of the map it is a remarkable monu-
ment in the history of cartography, and one which influenced
the mapping of many countries. It was not for half a century
after its initiation that a comparable undertaking was begun
in Britain by the Ordnance Survey. It was incidentally on
the initiative of Cassini de Thury that General Roy was commissioned to co-operate in the cross-Channel triangulation of
1787, and in so doing paved the way for the foundation of the
Ordnance Survey.

Having traced the history of the Cassini survey, we may
now examine the progress of general cartography in France.
J. D. Cassini's method of determining longitude was employed
at an early date in fixing positions outside France. Largely
with a view to improving existing marine charts, observers
from the last decades of the seventeenth century onwards
were sent to various countries in Europe, French Guiana, the
West Indies, Africa, and southern and eastern Asia, where in
course of time remarkably accurate values were obtained.
From Richer's observations, for example, the longitude of
Cayenne was determined within one degree of its true value.
The earlier results enabled J. D. Cassini in 1682 to sketch out
his famous planisphere, which incorporated forty determinations, on a floor of the Paris observatory. This was later
engraved with the title 'Planispherum terrestre', an issue of
which is known from 1694. These new observations were also
the basis of a collection of sea charts covering, on Mercator's
projection, the western coasts of Europe from Norway to
Spain; this was 'Le Neptune françois, ou Atlas nouveau des
cartes marines. . . . Revue et mis en ordre par les Sieurs Pene,
Cassini et autres', Paris, 1693.

The man who introduced this new work to the general
public, and in doing so effected what has been called the
'reformation of cartography', was Guillaume Delisle (1675–
1726). Guillaume was the son of Claude Delisle, a celebrated
teacher in his day of history and geography, to whom un-

doubtedly his son owed much instruction and assistance in his first ventures. The son also had the benefit of instruction in astronomy from Cassini at the Academy, of which he became an Associate in 1718. In 1700 Delisle began work as a compiler and publisher of maps, and for the rest of his life was the leader in cartographical progress with an international reputation. In his maps and globes he followed with understanding the progress of the Academy's work. Among his first productions was the 'Mappe-Monde Dressée sur les Observations de Mrs de l'Académie Royale des Sciences', 1700, a map in two hemispheres on the stereographic projection, carrying further the improvements of the 'Planispherum terrestre', and from time to time (e.g. 1724, 1745) amended versions were published. If compared with a modern map, the outlines of the continents are seen to be extremely accurate. Africa is particularly well drawn and correctly placed in latitude and longitude. South America is also well placed, though like North America, is still given too great an extent in longitude. The main area for which information is noticeably lacking is the northern Pacific, where Yezo (Hokkaido) is not yet clearly distinguished from the mainland, and ideas on the mythical 'Company Land' and the 'Strait of Anian' still plague the cartographer.

But if the continental outlines were now in great part known with considerable accuracy, the interiors of the continents outside Europe were still compounded of half truths, imagination and tradition. In dealing with them Delisle made another departure, for he was prepared to admit, by 'blanks on the map', the limitations of contemporary knowledge. In Africa, for example, he abandoned the system of central lakes which was an inheritance from the sixteenth century, and showed the main branch of the Nile as rising in Abyssinia, and elsewhere he displayed the same critical spirit. Since much information, in Asia particularly, still reposed on the authority of Greek and Latin writers, he gave much time and thought to determining the equivalents of ancient measures of length. As he lacked capital, and therefore the assistance of skilled engravers, Delisle's maps are not outstanding in their execution, but they are free of the mythical monsters and other devices with which the older cartographers had disguised their ignorance—or attracted their customers. In this respect, again, Delisle marks the transition to the modern map. His total output was not large, approximately 100 maps, in comparison with seventeenth-century map publishers, and much of his work was done to accompany works of travel or topography, for a map by Delisle was held to confer distinction on them.

He seems to have extended this simplicity of style to the representation of relief; he was certainly justified in objecting to some styles of mountain drawing, thought to enhance the attractiveness of a map, but on the main principle he was sound:

> "One of the main things demanded of a geographer is to mark clearly the rivers and mountains, because these are the natural bounds which never change, and which lead naturally to the discovery of geographical truths."

The improvement in the map of the world initiated by Delisle was continued and greatly extended by J. B. Bour-guignon d'Anville (1697–1782). His talent lay in the critical assessment and correlation of older topographical sources, and their reconciliation with contemporary observations. He was essentially a scholar, working mostly from written texts, which he collated with existing maps, and expressing his conclusions cartographically. Throughout his life he never journeyed beyond the environs of Paris. His extensive collection of cartographic material (10–12,000 pieces) was famous. Acquired by the French Government shortly before his death, it is now in the Bibliothèque Nationale, Paris. So great was his skill and industry that he soon acquired an international reputation as a map maker in an age in which classical scholarship still dominated the world of learning. D'Anville was in fact the last, and perhaps the greatest of those, who since the Renaissance, had followed this procedure, and he probably carried it as far as it was possible. He was one of the first to study the works of Oriental writers for details on the countries of the East. Further approach to accuracy could be attained only from exploration and actual survey of the continental interiors.

The first mark of recognition was bestowed upon him by

the Society of Jesus, when they entrusted him with the preparation for publication of the surveys of the provinces of
China, upon which members of the Order had been at work
since the later years of the seventeenth century. In many
instances these were based upon astronomical observations for
position, but in others were simply route surveys. From these
maps, Western Europe obtained the first reasonably accurate
and comprehensive conception of the geography of a large
part of eastern Asia. With the aid of these sectional surveys
D'Anville compiled a general map of the Empire of China.
The maps, forty-six in all on sixty-six sheets, accompanied
the 'Description géographique' of the Chinese Empire compiled by J. B. du Halde from the Jesuit reports, and were later
issued at Amsterdam with the title 'Nouvel Atlas de la Chine',
1737. An English edition of Du Halde with versions of the
maps appeared in 1738–41. D'Anville's share in this Atlas was
that of a compiler; but the efficiency of his general method of
work was displayed by his map of Italy, 1743, based upon a
critical study of Roman itineraries and measures of length.
The result was to reduce the area of the peninsula by "several
thousands of square leagues", and the accuracy of his deduc-
tions was strikingly confirmed by geodetic observations later
carried out in the States of the Church by order of Pope
Benedict XIV.

D'Anville's notable maps were those of the continents,
North America, 1746; South America, 1748; Africa, 1749;
Asia, 1751; Europe, in three sheets, 1754–60; and a general
map of the world in two hemispheres, 1761. The outlines and
positions of the continents, being based on the same data,
differed little from those of Delisle; their merit is displayed
in the treatment of the interiors. On the map of Africa, for
example, D'Anville went far beyond Delisle in removing the
conventional and largely fictitious topography, and his representation stood until the great journeys of the nineteenth
century inaugurated a new era in African cartography. D'Anville
took the correct view that the Blue Nile, rising in the Abyssinian
highlands, was not the principal branch of the Nile. Refusing
to break completely with Ptolemy's ideas, he depicted the main
river issuing from two lakes in the Mountains of the Moon, in

5° N. latitude and approximately 27° 30' E. longitude. The northward bend of the Niger is conspicuous, but is carried over 3° too far to the north, and the river truncated in the west. In the east it is connected with what may be intended for Lake Chad. In a note D'Anville states that there were reasons for presuming, contrary to common opinion, that the great river flowed from west to east. Elsewhere, except in the north, the detail is almost entirely confined to the coastal areas.

Another celebrated work was his map of India published in two sheets in 1752, the best map of the sub-continent before the work of Major James Rennell and the Survey of India.

D'Anville issued revised maps as the detail of contemporary exploration came to hand. In 1761 they were published as an atlas, and amended re-issues appeared until the early years of the nineteenth century. He paid great attention to draughts-manship and engraving—the lettering is clear and attractive— and in this respect his maps are greatly superior to those of Delisle, and to most of the products of his century. But perhaps his greatest contribution to cartography was due to the degree to which he carried out his own precept: "Détruire de fausses opinions, sans même aller plus loin, est un des moyens qui servent au progrès de nos connaissances".

D'Anville's work was carried on by his son-in-law, Phillippe Buache, who had a part in developing a more satisfactory method of representing relief on topographical maps, a problem which was receiving much attention at this time. On early engraved maps, hills and mountains, scarcely differentiated, were usually shown in profile, sometimes with shading to one side. These symbols are often called 'mole hills' or 'sugar loaves'. The decisive step was the advance from depicting ranges of hills or mountains as separate and isolated features to the representation of the surface configuration as an integrated whole. An interesting early example of this is the map of the upper Rhine valley in the Strasburg Ptolemy of 1513. On this the escarpments of the valleys are shaded, and the tributary valleys are incised in the uplands, which, however, are shown with a uniformly level surface. In countries such as Switzer-land, the first attempts were more in the nature of oblique perspective drawings; as the science of surveying developed, efforts were made to represent the actual area occupied by a range. This, combined with the profile, produced a three-dimensional effect.

One of the more successful methods of rendering relief was evolved by the Swiss cartographer, Hans Konrad Gyger (1599–1674).[1] In his maps of the Swiss cantons he attempted to show the land surface as though viewed from above, working in the folds and hollows by careful shading and leaving the higher areas untouched. His skilful workmanship, combined with his extensive personal knowledge of the country, produced a remarkably plastic effect, though he could convey only relative, not absolute, differences in altitude. That his method does not appear to have been generally followed was no doubt due to the lack of adequate data. For the next century at least the representation of relief was generally confined to shading valley slopes at a more or less uniform distance from the rivers. This style is employed, for instance, in the map of the environs of Paris made by members of the Academy of Sciences and engraved by La Pointe in 1678. Even on the sheets of the Cassini survey, seventy years later, no essential advance had been made, and the effect is much inferior to that achieved by Gyger.

The method of hachuring, by which relief is indicated by lines (hachures) running down the direction of greatest slope, may have been a development of this practice. The principle was fully worked out in the course of the eighteenth century to meet the requirements of military commanders. J. G. Lehman, on the analogy of the shadows thrown by an overhead light, propounded the theory that the greater the inclination of the surface to the horizon the heavier should be the hachuring, and he worked out a systematic scale for the thickness of the strokes. Hachuring, however, has several defects; if carried out elaborately, the heavy shading obscures much of the other detail on the map, and by itself can give no absolute value for the difference in elevation between one point and another. Moreover, without reference to other features, it is difficult to distinguish elevations from depressions.

[1]See Weiss, L., Die Schweiz auf alten Karten, 1945, pp. 107–66.

E*

The solution of the problem now generally employed is the contour line, i.e. a line running through all points at a given elevation. Unlike the hachure, it runs along, and not down, the slope. The origin of contouring is still somewhat obscure. An obvious contour is the line of high or low water, and it is not surprising therefore that it appears to have been developed in the Netherlands, at first to show the configuration of the sea bottom. Soundings off coasts and in estuaries are common on charts of the sixteenth century, on which banks are also enclosed by broken lines. It would not be a great step in advance to run these lines through soundings indicating a given depth of water. This appears to have been the practice by the beginning of the eighteenth century, by which time the number of soundings had greatly increased.

On a map of the Merwede estuary (1729), N. S. Cruquius, a Dutch engineer, showed depths by lines of equal soundings, referred to a common datum. Soon after, Philippe Buache drew a bathymetric map of the English Channel, with underwater contours at intervals of ten fathoms, but this map was not published in the Mémoires of the Paris Academy until 1752. In 1737 he had submitted to the Academy a chart of Fernao da Noronha with submarine contours, accompanied by a vertical section across an off-lying bank. Since he was also engaged in levelling operations in Paris, he must have recognized the applicability of the contour method to land surfaces. Its first use on land however is usually credited to Milet de Mureau, who about 1749 used lines of equal altitude on his plans of fortifications.

The eighteenth century was a period of great activity in the construction of canals, and it is therefore quite probable that the engineers responsible for them discovered the principle independently, just as Charles Hutton did in 1777 when seeking a method to determine the mass of Schiehallion, a mountain in Scotland. The general use in maps of large areas was delayed by the lack of sufficient data, though Cassini and others in France had calculated some heights by triangulation and by the barometer. The earliest British map to include spot heights appears to be Christopher Packe's 'Physico-chorographical chart' of Kent, 1743. Packe obtained his altitudes by the comparison of barometric readings.[1] Spot heights were frequently used before the end of the eighteenth century, e.g. on Mayer's 'Atlas de la Suisse', 1796–1802.

One of the earliest examples of the use of contours for a considerable area was Dupain-Triel's map, 'La France considerée dans les differentes hauteurs de ses plaines'. This purports to show France contoured at intervals of ten *toises* (about sixty feet), but the representation is largely influenced by his ideas on the orderly relations of mountains and plains. No general levelling had been carried out at this date, so that his contours were largely theoretical, but a number of summit heights are given, some of considerable accuracy, especially that of Mt. Blanc, and he added a vertical section across France. Dupain-Triel elaborated his methods and advocated their adoption in education in his 'Méthodes nouvelles de nivellement', 1802. Thus by the beginning of the nineteenth century the method was becoming known, and with the initiation of the great national surveys in the following decades it passed into general use.

A further step, the colouring of areas between successive contours by a given scale of tints, was taken in Stieler's 'Handatlas' of 1820. This hypsometric layering allows a general idea of the relief of a wide area to be formed rapidly. The value of contouring lies in the fact that, unlike hachuring, it enables the altitude of a particular point to be determined with considerable accuracy, for heights between contours can be estimated with practice. It does not, however, always permit an idea of the relief to be formed rapidly, and minor topographical features between contours are unrecorded. Consequently it is frequently combined with hachuring or hill-shading. In 1931, contours, hachuring, and layer colouring were all employed by the Ordnance Survey in the Fifth (Physical relief) edition of the One Inch map.

Note: With reference to the early use of contours, it should be noted that in 1697 Pierre Ancelin made a map of Rotterdam and the Nieuwe Maas which showed lines of equal depths.

[1]Campbell, Eila M., 'An English philosophico-chorographical chart.' (*Imago Mundi*, 6, 1949, 79.)

REFERENCES

CASSINI DE THURY, C. F., Description géometrique de la France. Paris, 1783.

DRAPEYRON, L., Enquête sur la première grande carte topographique, celle de France par C. F. Cassini de Thury. (*Rept. 7th Internat. Geogr. Congress*, 1899, vol. 2, pp. 897–920.)

FORDHAM, SIR H. G., Some notable surveyors and map-makers of the sixteenth, seventeenth and eighteenth centuries. Cambridge, 1929.

GOBLET, Y. M., France: L'évolution de la cartographie topographique. (*In* 'Catalogus mapparum . . .', Warsaw, 1933.)

PERRIER, G., Petit histoire de la géodesie. Paris, 1939.

SANDLER, C., Die Reformation der Kartographie um 1700. München, 1905.

WOLKENHAUER, W., J. B. Bourguignon d'Anville. (*Deut. Rundschau f. Geogr.*, v. 19, 1897.

CHAPTER X

THE BRITISH CONTRIBUTION IN THE EIGHTEENTH CENTURY

A DETAILED study of the British contribution to cartography before the eighteenth century lies outside the scope of this outline. Accounts of the achievements of men such as George Lily, Christopher Saxton, Norden, Speed, Ogilby and John Adams, to mention a few names only, can be found in the works of Sir George Fordham, Dr. Edward Lynam, and Prof. E. G. R. Taylor. In the second place, the emphasis here must be on the general development of maps and mapping, and it cannot be claimed that, however important in British cartography these sixteenth- and seventeenth-century cartographers are, they were in the van of technical progress. In the main they followed, often with a considerable time lag, the practice of their contemporaries in Portugal, Italy, the Low Countries and France. Saxton is surmised to have used methods of survey developed by Gemma Phrysius, and much of the attractiveness of his county maps is due to his Flemish engravers. The maps of Sanson and Delisle were industriously copied by English map publishers like William Berry, and the surveyors who in the eighteenth century won with their county maps the prizes offered by the Royal Society of Arts were not superior to the men who were producing the Cassini map in France.

There were of course exceptions to this generalization. The magnificent Molyneux globe of 1592, the first made in England and by an Englishman, was not excelled by any contemporary production. An important contribution to map projections was made by Edward Wright when he worked out mathematically the formula for Mercator's projection. The exiled Sir Robert Dudley was the first to employ this projection generally for the charts in his lavishly produced 'Arcano del Mare' (Florence, 1646). Nor must we overlook the stimulus which the epoch-making work of Newton exerted through astronomy and geodesy upon the development of cartography. Nevertheless, by and large, British map-making until the late eighteenth century was definitely behind that of other nations. Perhaps the best of a poor field in the first decades was Herman Moll, a Dutchman who came to London some time before 1682. His numerous maps are rather poorly designed and crudely engraved, but he made some effort to keep abreast of continental advances.

In 1738, John Green was lamenting the poor state into which the science had lapsed: he pointed out that cartography had fallen entirely into the hands of engravers, who copied each other without discrimination. Those "ignorant or mercenary Hands" who happened to become possessed of original material jealously concealed it from their rivals. To such conduct he attributed "the little Esteem, or rather great Contempt, that Maps are in here". He himself made some effort to remedy this state of affairs, in part as an employee of Thomas Jeffreys, but the first important British contribution was made through the development of instrumental equipment, which was effective in improving first the hydrographic charts, and then the maps.

In the eighteenth century the fundamental advances in mathematics and astronomy initiated by Sir Isaac Newton gradually bore fruit. The motions of the heavenly bodies were marked out, so that they could be accurately predicted for long periods, and eventually published annually in the 'Nautical Almanac' from about 1767. With the aid of lunar tables, the method of determining longitude within one degree by lunar distances was perfected. To this accuracy, the advances in the design of instruments also contributed; John Hadley had improved the quadrant by the introduction of reflecting mirrors[1], and more accurate readings were obtained from the use of the vernier scale. Meanwhile, John Harrison was engaged in designing and constructing a time piece which would be sufficiently robust and accurate to allow longitude to be determined from the difference between local time and the time indicated by the chronometer for a given meridian.

[1]The method had earlier been described by Sir Isaac Newton.

Harrison was eventually awarded the prize offered by Parliament to "such person or persons as shall discover the Longitude" in 1772, and a copy of his successful chronometer was used by Capt. Cook on his second and third voyages, giving extremely accurate results. This method by 'transport of chronometers' finally superseded that by lunar distances.

Though these instruments were initially used in navigation and hydrographic survey, it must be remembered that the explorers of the following century relied largely upon the sextant (an improvement on the quadrant) and the chronometer for the surveys they were able to accomplish. Another survey instrument which at this time emerged in its essentials was the theodolite, a descendant of the 'polymetrum' devised in the early sixteenth century. By the invention in 1763 of his graduating engine, Jesse Ramsden solved the problem of dividing the brass circle accurately, and then worked on his famous theodolite for a number of years. This included a horizontal circle three feet in diameter, which by the aid of micrometers enabled readings to single seconds to be obtained. The sighting vane of the older models was replaced by a telescope moving freely in the vertical plane of the instrument. This theodolite was admittedly heavy and cumbrous, but it proved the most efficient instrument for observing angles in survey, and by gradual modifications it has developed into the highly precise and portable patterns of today. This instrument was first employed in the connexion by triangulation of England and France in 1787, and later in the Ordnance Survey of Britain and in India.

The first results of these technical advances were seen in the increased accuracy of hydrographic charts, and in their production and publication Great Britain assumed the lead which she has maintained for the last 150 years. The end of the eighteenth century may be approximately taken as the point at which the general outline of the continents, outside the Polar circles, and their precise position had finally been determined, though much patient and careful work had still to be carried out before all the details were filled in. In the remainder of this outline, therefore, we shall no longer be concerned with the seaman's contribution to the 'unrolling of the map', but must leave it with a brief tribute to the work of Cook, Vancouver, Flinders and their colleagues in Pacific, Australian and Antarctic waters, and to their successors, Fitzroy, W. F. Owen, P. P. King, Moresby, Nares and other distinguished navigators. Upon the work of these men are founded the charts issued by the Hydrographic Department of the Admiralty.

Two extra-European fields for cartographic work were open to British surveyors in the eighteenth century, North America and the Indian sub-continent, and in both they acquitted themselves creditably, not only paving the way for subsequent advances, but providing the first adequate maps of those areas. For North America, apart from the coasts and the immediate hinterland in the east, only maps based on the rough sketches and reports of explorers were available before the middle of the century. One contemporary cartographer was candid enough to admit that beyond the Great Lakes, the detail was 'in great measure guess work'. The progress of settlement, the organization of the colonies, and particularly Anglo-French rivalry, created a demand for general maps of greater reliability, and led land surveyors to turn from their work on estates and plantations and to apply themselves to the wider problem. In the early years, considerable encouragement was given by the Lords Commissioners for Trade and Plantations. Two notable general maps incorporate the results of this activity. In 1749, Lewis Evans published his 'Map of Pensilvania, New-Jersey, New-York', etc., on the scale of 20 miles to 1 inch, based on numerous determinations of latitude, and two longitudes, those of Philadelphia and Boston. To these, he had fitted the 'Draughts or Discoveries', with which many gentlemen had furnished him. That the map was based in great part on route surveys by distance and bearing is shown by his remark "No distance could be taken but by actual Mensuration (the Woods being yet so thick)", i.e. the surveyors were unable to triangulate with the circumferentor or early theodolite. Six years later, Evans issued his most important map, the 'General map of the Middle British Colonies in America' (1 inch to 50 miles). This map was in great demand, and was much used in North America during the Seven Years' War. When Governor Thomas Pownall, himself something of a surveyor, issued another edition in 1776, he could say, "Where local Precision has been necessary this Map has been referred to not simply in private but public Transactions, such as the Great Indian Purchase and Cession".

Governor Pownall was also associated with another and more famous map of this period; John Mitchell's 'Map of the British and French Dominions in North America', published by Thomas Jeffreys in 1755. Mitchell was a botanist who had settled in Virginia early in the century, returning to England in 1747. Little is known of his cartographical work, and his map probably owes something to Jeffreys. It represents eastern North America from the southern shores of Hudson Bay to the Mississippi delta, on the scale of approx. forty-three miles to the inch. A prominent feature is the detailed statement of the authorities followed in its compilation. It is stated that the map was undertaken

"with the Approbation and at the request of the Lords Commissioners for Trade and Plantations and is chiefly composed from Draughts, Charts and Actual Surveys . . . great part of which have been lately taken by their Lordships' Orders".

With this official character, it is not surprising that the map played an important part in the peace negotiations between the American colonies and Britain in 1782, for on it the boundary between Canada and the United States was laid down.

The second great field for British cartography was in India. Before the time of D'Anville there was nothing approaching an accurate map of the sub-continent, and large areas of the interior were blank. From about 1750 onwards, the East India Company, to forward their commercial expansion, actively promoted the charting of the coasts, and this work was later stimulated by their hydrographer, Alexander Dalrymple. The first systematic land surveys resulted from the activities of Major James Rennell in the Bengal Presidency, the first extensive area to fall under the Company's complete control. In his twelve years in India (from 1767 to 1777 he was the first Surveyor-General of Bengal) Rennell initiated and directed a comprehensive and uniform survey of Bengal and Bihar, to meet military, administrative and commercial demands. The survey was based upon a network of distance and bearing traverses, controlled as the work progressed by cross bearings and closed circuits. A further control was afforded by observations for latitudes. Distances on large-scale work were measured by chaining, in other cases by perambulator. Quadrants were employed for horizontal angular measurements as well as for obtaining latitudes, and theodolites were gradually introduced. Much of the work was based on traverses along the rivers and main roads, details of the countryside being largely filled in by estimation. Given the methods employed and the difficulties encountered—Rennell himself was severely wounded and suffered constant attacks of fever— the results were extremely creditable, and the standard of mapping was much higher than that of many European countries.

The first edition of the 'Bengal Atlas' ,with maps on a scale of 5 miles to 1 inch, was published in London in 1779, two years after Rennell's return on pension. In London, he continued to maintain his interest in the mapping of India, and in 1782 published his great 'Map of Hindoustan' with a 'Memoir'. This map, in four sheets on the scale of one equatorial degree to one inch, was a remarkable piece of compilation. The sources used are critically discussed in the 'Memoir'; they included the East India Company charts communicated by Dalrymple and route surveys by military engineers accompanying military expeditions, adjusted to an astronomical framework of latitudes and longitudes, the latter mainly for coastal cities obtained from the eclipses of Jupiter's satellites. For the Punjab he relied largely upon a map done by a native, giving the courses and names of the five rivers, "which we have never had before". With the advance of British arms, Rennell was continually receiving fresh material, and six years later he published a revised and enlarged edition (1½ in. to 1 degree), and this was followed by further editions in 1792 and 1793.

Knowledge was now increasing so rapidly that it became impossible to compile a map of the whole country on the methods employed by Rennell; however, until the emergence of an organized Survey department and the completion of the Great Trigonometrical Survey in the next century, Rennell's 'Hindoustan' remained the basis of Indian cartography.

All this activity on land and sea was making available a great mass of cartographic material to the map publishing houses in London. It was through the output of these firms, in which the new facts were collated and presented in convenient format, that the work of surveyors all over the world ultimately reached the public. No longer did this filter at second hand through the publications of continental establishments. London had become the universal centre of cartographic progress. During the period of the revolutionary and Napoleonic wars, the seas were the virtual preserve of British sailors, and maritime, commercial and military enterprises, while requiring the best available maps and charts for their execution, provided in return a mass of observations and records by which the existing material could constantly be amended. British cartographers availed themselves to the full of these opportunities, and for the first time their work received international recognition. Parallel with this expansion, there was a marked improvement in the construction and engraving of their maps, which, by their clarity and freedom from conjectures or unverified detail, in themselves conveyed a general impression of accuracy and thoroughness. British cartography was thus freed from its dependence upon continental sources.

The beginning of this advance is to be found in the work of Thomas Jeffreys. He was the publisher of Benjamin Donn's one inch to the mile map of Devonshire, the first county map to win the award of £100 offered by the Royal Society of Arts, 1765, and himself surveyed several counties. His most important later work was the publication of the improved charts of the American coasts resulting from the labours of men like James Cook. Important collections of these—American Atlas, North American Pilot, and West Indian Atlas—were published after his death by his successor, William Faden. At Faden's establishment the first sheets of the Ordnance Survey maps were engraved, until that department secured a staff and offices of its own.

Contemporary with Faden was John Cary, who maintained a high standard of excellence in his maps and globes. Both these men gave much attention to preparing new maps of the British Isles, utilizing the numerous county surveys, and later the early editions of the Ordnance Survey sheets, besides adding much detail from their own work. Cary in particular paid attention to the rapidly developing system of communications; in 1794, he was engaged by the Postmaster General to supervise the survey of some nine thousand miles of turnpike roads in Britain. The results were incorporated in various road books and county atlases, the latest and largest of which was the folio 'New English Atlas' of 1809.

But the man who established the international reputation of British cartography was undoubtedly Aaron Arrowsmith. A native of Winston, Durham, Arrowsmith was typical of the generation which gave Britain its lead in the technical revolution of the eighteenth century. With no advantages of birth, or systematic education, he acquired a knowledge of mathematics, and of the theories of map projections, and through a long apprenticeship became proficient in the practical technique of map production. He came to London in 1770, and worked for some time as a land surveyor; as such he is described on Cary's 'Map of the Great Post-Roads between London and Falmouth' of 1784, for which he was largely responsible. It is possible that it was in Cary's establishment that he learned the technique of map engraving. However that may be, he set up for himself as a cartographer and map publisher some time before April 1790. His first publication, a chart of the world on Mercator's projection, which when mounted had the considerable dimensions of 5 ft. by 8 ft. 4 in., was an immediate success. This included the tracks of the most important navigators since the year 1700, all "regulated from the accurate astronomical observations" taken on the three voyages of Capt. James Cook.

It was his proficiency in 'regulating' observations from varied sources and in fitting together sketch maps or reports by numerous explorers and travellers that gave Arrowsmith his pre-eminence. His fame also owed something to the style of engraving. The names are clearly engraved, and much detail is given without confusion. Save for the title cartouches, the maps are entirely without decorative details, behind which lack of knowledge so often had taken refuge. On his maps in general, relief is poorly represented; he considered that altitudes could not be introduced except on very large scales. Four years later, this map, which in his words met with "great approbation", was followed by another of the world on the globular projection, published with a 'Companion', in which he set forth his opinion that the Mercator and globular projections were the most suitable on which to represent the whole surface of the globe. For this second map, he had corrected the positions of some hundreds of places, and considered that "as far as the name can apply to a map", it was "an original work". In the 'Companion', Arrowsmith lists nearly 140 authorities on which he had drawn, including a number of manuscript maps of the Hudson's Bay Company's territories by Philip Turnour, astronomical observations by Cook's officers, and three maps of the country north of Fort Churchill by an Indian. Alexander Dalrymple had also presented him with a complete set of his geographical publications, including 623 maps and charts.

It is not possible to list here all Arrowsmith's productions, but notable among them are his chart in nine sheets of the Pacific Ocean, 1798, the dimensions of which when mounted are over 6 ft. by 7 ft. 6 in., and which is now a valuable source for the history of Pacific exploration; his maps, nineteen sheets in all, published in conjunction with Thompson's 'Alcedo; or dictionary of America and West Indies', and based on original materials "that have till lately remained inaccessible at Madrid and at Lisbon".

After his death in 1825, his business was carried on for a time by his sons Aaron and Samuel, but was later taken over by his nephew, John (1790–1873), who maintained his uncle's reputation. He was closely in touch with the explorers of Australia, working up and publishing their maps. Abandoning his uncle's practice of issuing large maps suitable for mounting as wall maps, which had alternatively to be bound somewhat inconveniently in sheets, he worked for a number of years on the sheets of an atlas, uniform in size and style. When published in 1840 as 'The London Atlas of universal geography' it was the best of its kind as a political and location atlas.

REFERENCES

ARROWSMITH, A., Companion to map of the world on the globular projection. 1794.

CRONE, G. R., John Green, a neglected eighteenth-century geographer and cartographer. (*Imago Mundi*, 6, 1949, 85.)

FORDHAM, SIR H. G., John Cary; engraver, map, chart and print-seller and globe-maker, 1754–1835. Cambridge, 1925.

GIPSON, L. H., Lewis Evans, Philadelphia, 1939.

GOULD, R. T., The marine chronometer; its history and development, 1923.

PHILLIMORE, R. H., Historical records of the Survey of India, Vol. 1. Dehra Dun, 1945.

RENNELL, J., Memoir of a map of Hindoostan, 1783.

SKELTON, R. A., Captain James Cook as a hydrographer. (*Mariner's Mirror*, 40, 1954, 92–119.)

STEVENS, H. N., Lewis Evans. His map of the British Middle Colonies in America. 2nd ed. 1920.

CHAPTER XI

NATIONAL SURVEYS AND MODERN ATLASES

CARTOGRAPHY since the early decades of the nineteenth century is characterized by the execution of regular topographical surveys as national undertakings. Most has been accomplished in Europe, in some countries of Asia (e.g. India, Japan, the Dutch East Indies); in the United States and Canada; and in Egypt and parts of North Africa. Though similar surveys have been begun elsewhere, progress has not been rapid, and great areas of the earth's surface are still unmapped at medium scales on a systematic trigonometrical framework. For these the cartographer depends upon miscellaneous and unco-ordinated material of varying quality, produced by travellers, boundary commissions, railway and road development, settlement schemes, and mining and similar concessions. To these must now be added rapid reconnaissance surveys, mainly from the air, of considerable areas carried out during the last war.

The second major advance has been in the enlarged scope of atlases, and the increasing use of mapping as a technique in dealing with a wide variety of problems in physical and human geography, and in administration. This progress was considerably assisted by the change from engraving on copper plates to colour lithography and its modern developments, which allow a great variety of detail to be clearly shown.

The great national surveys of the nineteenth century rested upon methods resembling in general those of the Cassinis. These were gradually refined as instrumental design progressed, and corrections were applied to the observations to allow for factors previously neglected. These included corrections for refraction and the curvature of the earth's surface, for changes in temperature and other conditions affecting the measurement of base lines by metal tapes, and for the reduction of the standards of length employed in the field to the official national standard. By the further careful comparison of national standards such as the yard and the metre, it was possible to determine, from the results of surveys in many parts of the world, the figure of the earth with considerable accuracy. The shape of the earth approximates to a spheroid, flattened at the Poles; the International Union of Geodesy in 1924 adopted a figure for the major semi-axis of 6,378,388 metres and of 1 in 297 for the flattening. Correction for the figure of the earth is of course essential in the computation of the triangles.

It should be noted that maps are drawn as though projected on to the plane of the sea-level surface, so that distances measured on them are independent of irregularities in the relief.

The stages in a systematic topographical survey, before the introduction of aerial photography, may be briefly summarized as: (i) determination of mean sea level at one point at least, to which all altitudes are referred; (ii) a preliminary plane table reconnaissance to select suitable points for the triangulation, and the erection of beacons over them; (iii) determination of initial latitude, longitude and azimuth (for direction), which will 'tie' the map to the earth's surface; (iv) careful measurement of the base or bases with a tape or wire of a special alloy; (v) triangulation, the theodolite being used to observe horizontal angles from the base and beaconed points, and to measure altitudes by readings of vertical angles; (vi) calculation of the triangulation and heights, and the transference of the trig. points to the sheets issued to the plane tablers; (vii) the filling-in on the sheets by the plane tablers of the required topographical detail—contour lines, rivers, woods, settlements, routes, and names.

In the present century, the measurement of long meridian arcs has tended to go out of fashion, partly because they of necessity take no account of local topographical features and consequently may not be particularly useful for local surveys. In areas of excessive local gravity anomalies, they do not furnish the expected data for determining the figure of the earth. On the whole, methods of survey have become much more flexible, and are adapted to local conditions. It is found that "it is better to be content with small triangles easily accessible than to make enormous efforts at rays longer than nature easily allows". With improved equipment, the necessary base lines can be measured rapidly and accurately.

Instrumental development has also given the surveyor greater freedom. In place of the old cumbrous theodolites, light and accurate instruments have been evolved, such as the 3½-inch 'Tavistock' theodolite, which allows the mean of readings on each side of the glass circle to be read directly through the same microscope to one second of arc. The invention of wireless has also simplified the troublesome problem of determining longitudes. It is relatively simple to receive Greenwich mean time by time signal and to compare it with local time. The development which has perhaps attracted most public attention has been that of air survey, though often its merits have been exaggerated. As long ago as 1858, the value of vertical air photographs, taken from balloons, was appreciated, but there were obvious difficulties in obtaining them. A combination of the camera and theodolite was later successfully used in ground survey, particularly in Canada. Experience of air photography gained in the first world war gave a considerable impetus to research into air survey, and by the end of the last war, owing to the demand for the rapid mapping of territories inaccessible to land surveyors, standard methods had been evolved.

The problems of air survey are concerned with (1) obtaining suitable photographs, (2) providing the necessary ground control for the framework of the map, and (3) filling in the detail from the photographs. First, the area to be mapped must be covered by overlapping strips of photographs, taken at a constant altitude and in favourable conditions. In England there are on the average only thirty days a year suitable for air photography. These photographs are examined stereoscopically in pairs. Vertical, or nearly vertical, photographs simplify the later stages of the work. To produce a map on a relatively small scale and covering an area with slight surface variations, the centres of photographs with a tilt of not more than 2° can be treated as plane table stations and rays drawn from them to prominent features portrayed. Details obtained thus are tied in to a relatively small number of points fixed by triangulation on the ground.

For maps on the larger scales required, for example, for civil engineering projects, much more precise results are obtained from overlapping stereoscopic pairs of photographs in a plotting machine. This operation depends on complex optical principles, but stated very simply it involves placing the photographs in their exact relationship to each other and to the ground surface (thus eliminating tilt). The operator, viewing them stereoscopically, and thus having before him a three-dimensional representation of the surface, is enabled to trace its features, including contours, by the intricate mechanism of the plotting machine.

To reduce the number of ground control points required, radar navigational aids can be employed to fix the position of each photographic exposure with sufficient accuracy. Professor Hart quotes instances in which areas inaccessible on the ground have been mapped on scales as large as 1:50,000 from control stations 250 miles distant. In such cases, however, difficulties in contouring may arise from lack of data. To fill in the detail requires considerable practice in the interpretation of the photographs. Types of soils, rock formation, and vegetation, for example, will reveal themselves to the practised eye. Their appearance naturally will change under varying conditions of light, and, indeed, for certain purposes, the photographs must be taken at a certain time of day or season of the year. The methods and standards of air survey, and hence the expenditure involved, can within limits be varied according to the accuracy required.

The 1939–45 war gave a great impetus to air survey. Under the direction of the U.S. Aeronautical Chart Service, for example, some 15,000,000 square miles, equivalent to more than a quarter of the land surface of the earth, were photographed from the air by trimetrogon cameras (multiple lens cameras) for small-scale mapping. It has been said that "In the field the aerial camera achieved its final triumph over the plane table as the cartographic surveyor's primary tool for picturing the earth's face for mapping purposes".[1]

The survey of a country by the methods described is clearly a costly task, requiring a large highly trained field staff, not to mention the establishment necessary to compile, draw, and print the maps for public issue. A further cause of heavy expenditure, in industrialized countries especially, is the need for constant revision. It is not surprising therefore that progress in mapping has been slow in under-developed countries.

The Ordnance Survey

The history of the Ordnance Survey of Great Britain illustrates the problems encountered in the development of a national series of topographical maps, and the extent to which they have been influenced by the varying requirements of their users.

The Ordnance Survey (at first known as the Trigonometrical Survey) was officially established in 1791, being the outcome of survey operations for the connexion of England and France by Cassini and William Roy in 1787. In its early days the Survey had two tasks, the carrying out of the great triangulation between 1798 and 1853, and the production of the One Inch to a mile map. The triangulation rested on two base lines, one on the shores of Lough Foyle, the other on Salisbury Plain, measured respectively in 1827 and 1849. When a test base was measured at Lossiemouth in 1909, it was found that the error on any side of this triangulation did not exceed the order of one inch in a mile. A new primary triangulation, for which some of the original stations were used, was carried out in 1936–38, and again revealed the accuracy of the old work. As the triangles were carried across country, the work of the One Inch survey proceeded. The first four sheets, issued in 1801, covered Kent and part of Essex and London. The purpose of this map was largely military, the scale being convenient for the movement of infantry. It was not until 1870 that it covered the whole of Great Britain.

[1]Wright, J. K., in Comptes rendus, Congrès Internat. de Géogr., Lisbon, 1949, I.304.

Meanwhile the land question in Ireland had created a demand for maps on a larger scale which would allow the areas of smaller administrative units to be shown clearly. The six inch to one mile survey of that country was consequently begun in 1824. Later, survey on this scale was extended to Great Britain, and from 1840 the One Inch sheets of northern England and Scotland were reductions from the Six Inch. The latter is now the largest scale which gives complete coverage for the whole country.

With the industrial developments and the great expansion of towns and communications in the mid-nineteenth century, the demand for large-scale plans became insistent. In 1858, partly under the influence of continental ideas, it was decided to publish plans of the whole cultivated area on the scale of 1:2,500 (as it happens, this is very nearly equivalent to twenty-five inches to one mile). The twenty-five inch is now the base from which, so far as it extends, all smaller-scale maps are derived.

For many years, apart from what were essentially index maps to the various series, the One Inch to a mile was the smallest scale map published by the Ordnance Survey. It was not until 1888 that the Quarter Inch map was completed, though it had been started in 1859 at the instance of the War Office and the Geological Survey. This was followed some twenty years later by the Half Inch and the Ten Mile maps. The demand for these in the first place was mainly military, but with the development of the motor car they have become increasingly popular.

Recently, another map has been added to the national series, the 1:25,000 (approximately two and a half inches to the mile), begun in 1945. This fills usefully the gap between the One Inch and the Six Inch and is a scale commonly in use on the Continent. This is the smallest scale on which it is possible to show roads and similar features without having to exaggerate them for clarity, and to include most minor topographical features. Up to date, sheets covering most of England and Wales, except for central Wales, and parts of Scotland have been published in a 'provisional' edition. Here provisional indicates merely that it is based on the existing Six Inch survey.

The regular edition will be derived from the re-survey of Great Britain. Contours at twenty-five feet interval will be from ground survey or from air photographs. On the provisional edition, intermediate contours are interpolated.

Since the methods of representing surface forms have been closely related to printing techniques, the two may conveniently be considered together. The first edition of the One Inch was printed from engraved copper plates, and relief was somewhat crudely shown by hachures, very much in the style of the Cassini map of France. At first, the heavy hachuring tended to obscure detail, but later there was some improvement. Contours were first adopted about 1830 as a result of experience on the Irish Six Inch survey, and were soon afterwards introduced on the Six Inch and the One Inch maps of northern England which completed the first edition. The contours were surveyed instrumentally at 50 feet, 100 feet and thence at intervals of 100 feet to 1,000 feet. Above 1,000 feet, the interval was 250 feet. Considerably later, on the Popular (Fourth edition) of the One Inch map, additional contours were interpolated at intervals of fifty feet.

Since all the detail on the engraved sheets was in black, the contour lines were not always conspicuous. Though hachures in brown were later printed on the One Inch from a second copper plate, the use of colour in general followed the introduction of lithographic printing or a development of it, photozincography, in which the original was photographed and transferred to zinc plates for printing. In the Third edition of the One Inch, completed in 1912, and known as the 'fully coloured', relief was shown by hachures in brown and contours in red. In all there were six printings, brown and red for the relief, blue for water, green for woods, and burnt sienna for roads, with names and other detail in black. Following Bartholomew's successful production of a Quarter Inch map of Britain in which relief was shown by layer colouring, the Ordnance Survey employed this method on its Half Inch map, produced at the beginning of this century, and also for various district maps.

In the years before 1914, experiments in the best methods of showing relief were carried on vigorously. One of the most beautiful and satisfactory of these was the One Inch sheet of Killarney, for which no less than thirteen separate printings were employed, a fact which, even in those days of relatively cheap costs, prohibited its general adoption. In this sheet, all the earlier methods were combined: contours in black dotted lines, hachures in brown, hill shading by heavier hachures on the south-eastern slopes (giving the effect of lighting from the north-west corner of the map), and delicate layer colouring. The sheet gives a most expressive representation of the modelling of the relief, even in the less elevated areas.

In the interval between the two wars, it became necessary, owing to the state of the old copper plates, to re-draw the One Inch afresh, and the opportunity was taken to introduce several improvements.[1] An entirely new style of lettering, based on that of Trajan's Column, was introduced, adding much to the legibility and appearance of the map. This Fifth (Relief) edition also incorporated in a modified form some of the features of the Killarney sheet: relief was shown by contours in brown, hachures in orange, hill-shading in grey with layers in tints of buff. Again, the effect of modelling is very expressive, but, unfortunately, it was considered that this style was not popular with the public and the Relief edition was abandoned for a less elaborate style. The present Sixth (New Popular) edition follows closely the Fifth, though relief is shown by contours only, conspicuous in brown. The sheets have been cleared of some detail, including the black symbol which differentiated between woods of deciduous and coniferous trees.

In the course of 150 years, the sheet lines have undergone considerable changes, which have been related to the projection and central meridians employed. In the early days of the First edition, the One Inch sheets of northern and southern England had separate central meridians, and the large-scale county plans were also drawn on their own meridians; consequently it was impossible to fit sheets of adjoining counties together. The size of the sheets has also varied from time to time. With the re-calculation of the whole survey on one pro-

[1]Previously, for each revision, the outline of the engraved copper plates, after correction, had been transferred to stone or zinc for reproduction.

jection, the Transverse Mercator with the central meridian 2° W., and the introduction of the National Grid, uniformity has now been secured. The grid serves two purposes; for the smaller-scale maps it provides a national system of reference, by which a point can be located by an identical reference on maps of all scales; secondly, on the larger-scale plans, it provides data by which, with certain corrections, very large-scale survey, such as is done by civil engineers, mining surveyors and the like, can be carried out accurately.

The National Grid is based on the central meridian, and divides the country up, in the first place, into squares with one-hundred kilometre sides. For convenience in numbering, the origin of the grid is placed a little to the south-west of the Scilly Isles. These squares in turn are divided into 10 km. and kilometre squares, which are shown on the 1 : 25,000 and One Inch maps, while the Six Inch map has the kilometre squares. By estimation, therefore, references to the nearest 100 metres can be read from the One Inch.

With the introduction of the National Grid, the sheets of the plans and maps are now regularly arranged; for example, 100 sheets of the twenty-five inch plan form one sheet of the 1:25,000 map, which in turn covers one 10 km. square on the One Inch map. Thus the smaller-scale map in each case forms an index to the larger. The One Inch map, however, is not designed on regular sheet lines. Great Britain has a long and much indented coastline, and the adoption of a regular system would inevitably lead to the production of sheets showing very little land surface. Consequently the sheets have been 'fitted to the topography'. Their size, however, has now been standardized, covering 45 km. from north to south, and 40 km. from east to west. The effect of this has been to reduce the number of sheets (the Sixth edition covers England and Wales in 115, in contrast to the 146 sheets of the Fifth), and, by allowing generous overlaps, to reduce the necessity for special District maps.

With the carrying-out of recommendations of the Departmental Committee of 1938, Great Britain now has a national series of maps and plans designed on a common system, which is not surpassed by those of any other country. No other national survey publishes a regular series on the scale of 1 : 2,500. One may perhaps regret that the contour is the sole method employed, except on the smallest scales, to represent relief. Minor topographical features of local importance, occurring between contours, are necessarily omitted, but could be indicated by hachuring or shading; very often a significant crest line, for example, is not brought out by contouring. Contours, however, do not obscure other detail, and in strongly accidented relief convey some visual impression. They are certainly essential on modern medium- and small-scale maps; supplemented by a restrained use of other methods, they would supply the best solution. Judged by clarity of detail, use of colour, lettering, and general design the Ordnance Survey maps set a very high standard.

Outside the British Isles, this country has been responsible for the mapping of the colonial territories, and has contributed in personnel and technique to that of India in particular, but also to other countries including Egypt and Siam. For many years, regular surveys were not undertaken in the territories, and large sums were expended on maps for particular purposes as required. It is now acknowledge that a regular topographical survey, far from being a luxury, is an essential preliminary to sound development, and therefore an economy in the long run. This was officially recognized by the establishment of the Colonial Survey Committee in 1905, which concentrated mainly on tropical Africa. Most colonies now have their own survey departments, but since they were at first entirely dependent on local finances, their progress was necessarily hampered, especially after 1931. The establishment of the Directorate of Colonial Surveys in 1946, ensured a central direction and more adequate resources. The task before the Directorate is a heavy one; when it was set up some 1,500,000 square miles had still to be mapped, though a proportion of this total is not an urgent necessity. Air survey is well adapted to work of this type, and the Directorate is making considerable use of it.

Some Foreign Map Series

It will be useful for comparison to note briefly what has

been done in France, and to glance at the practice of other countries in representing relief where it presents greater difficulties than in Britain. In France the national survey is the responsibility of the Institut Géographique National, the civilian body which has replaced the Service Géographique de l'Armée. The two main topographical scales are 1:20,000 and 1:50,000, but neither of the current editions give complete coverage.

In the nineteenth century, the Cassini series was replaced by the Carte de l'Etat Major on the scale of 1:80,000. Originally engraved, later photolithographed, this map has no contours, but is heavily hachured. About 1900, a new series on the scale of 1:50,000 was initiated. The original elaborate scheme has been considerably modified, but progress has been slow. Some 220 sheets, out of a total of about 1,100, have been published; owing to French pre-occupation with their eastern frontiers most of the sheets fall in this zone. In fact the only medium-scale map completely covering France is an enlargement of the old 1:80,000 to 1:50,000.

Since the reorganization and with the use of air photographs and stereoplotting machines, progress has been somewhat more rapid. The 'Nouvelle Carte de la France au 50,000 ème' is contoured in brown (black on rocks and blue on glaciers), with contours at 5, 10 or 20 metre intervals, according to the character of the ground. Relief is further emphasized by hill shading both 'vertical' and oblique from the north-west to impart a plastic impression. This is obtained by photographing a relief model suitably illuminated. Considerable attention is paid to the vegetation cover; various types are indicated by symbols in black with a flat green overprint.

The national maps of Switzerland are distinguished by the high standard achieved in the representation of relief, the result of long experience and experiment. In 1938, the well-known Siegfried map, which in its day set a good standard, was replaced by a new series on the scale of 1:50,000, more accurate and legible. It is very closely contoured, at 20-metre intervals, with less obtrusive subsidiary contours at 5 and 10 metres in brown, black or blue according to the surface. Summits, precipitous slopes, rock falls and similar features are indicated by fine rock drawing in black. Glaciers have contours in

F

blue and a light blue tint, with moraines in brown. Relief is further emphasized by shading in a neutral tint or blue-grey put in by the draughtsman, not by photography as in France. The general effect is most expressive, largely due to the careful rock drawing. The other details have not been allowed to dominate the relief, the green undertone of the woods, for example, is light, and the conventional signs are neatly drawn. A new series on the scale of 1:20,000 is also in preparation. When the two are complete, Switzerland will be unrivalled in its standard of national cartography.

In the United States, there is no single official mapping authority as in Great Britain. The U.S. Geological Survey is now the chief agency for topographical mapping, but a number of others produce maps and charts for special purposes. For most of the nineteenth century, the principal demand, as the tide of migrants flowed westwards, was for the rapid survey of the vast tracts of land to facilitate settlement. From 1776, this was carried out for the central government by the General Land Office. At the same time, individual States produced small-scale maps of their areas, generally of no great order of accuracy.

More precise surveys were gradually carried out by other bodies. The Coast Survey (now the Coast and Geodetic Survey) was established in 1807, but accomplished little for its first thirty years. In addition to its chief task of charting the coasts and mapping contiguous land areas, it is responsible for the basic network of triangulation and levelling used in other surveys. The Corps of Topographical Engineers, founded a few years later, was at first largely engaged in meeting military requirements and in exploration, but was afterwards employed in tasks connected with the improvement of rivers and harbours, in the survey of the northern lakes and in boundary work. It also carried out much survey in the territories west of the Rocky Mountains around the middle of the century. In 1879, to co-ordinate this work, all surveys, geological and topographical, west of the 100th meridian were entrusted to the newly established Geological Survey, whose sphere of activity was eventually extended over the whole country. It is now responsible for eighty per cent of the topographical survey performed by government agencies.

The topographical maps issued to the public are on three standard scales, according to the importance of the particular area; 1:31,680; 1:62,500; and 1:125,000. The sheet lines are based on quadrangles formed by parallels of latitude and meridians of longitude, the sheets of the 1:62,500 series covering 15' of latitude and longitude. At this scale the contour interval ranges from 10 to 50 feet. Every fourth or fifth contour line is strengthened, and numerous spot heights and bench marks are given. The sheets are printed in three colours; cultural features (roads, settlements, etc.) and names in black, water features in blue, and contours in brown. On some, a green tint is used for woodlands. The general effect is clean and sharp, with no over-crowding of detail.

About twenty-five per cent of the area of the United States is covered by 'acceptable topographic maps', but for almost forty per cent of the area no topographic maps of any kind exist. Recently, however, progress has been expedited, largely by the adoption of air survey methods. Since 1936, the Geological Survey has made increasing use of air photographs and carried out extensive research in techniques and instruments. Photographic cover of varying standards exists for all but five per cent of the country, and is used for many administrative purposes. It is anticipated that within twenty years all the United States and Alaska will be covered by standard topographic maps. The Survey has also produced geological maps on the scale of 1:62,500, or larger, for ten per cent of the country.

The International Map of the World on the Scale of 1/1 *Million*

The value of a map of the world on a uniform projection and scale with a standard set of conventions to many types of map users is obvious, but equally obviously its production on any scale larger than those employed in atlases could not usefully be contemplated until a significant proportion of the earth's surface had been mapped topographically. The idea of such a map was first advanced by Professor Albrecht Penck at the International Geographical Congress, Berne, 1891, when he proposed that it should be compiled on the scale of 1/1 Million (approximately 1 inch to 15.8 miles). Little was achieved for twenty years, when the British government invited foreign delegates to a conference in London, at which the 'Carte internationale du Monde au Millionième' was initiated on an approved system. The projection is a modified polyconic, which allows adjoining sheets to be fitted together, each covering 4° of latitude, and 6° of longitude, though nearer the Poles than latitude 60° two sheets may be combined. Relief is by contours, generally at 100-metre intervals, and layer colouring, with shading for minor features, according to an approved pattern. Each national survey is responsible for the sheets covering its own territory, and names are given in the local form.

From the beginning the project encountered difficulties. It is clearly almost impossible to devise a scheme of contour intervals and layer colouring which will depict satisfactorily all kinds of topography, from the Himalaya and the plateau of southern Africa to the English Plain, and in practice consider-able latitude in the selection of contours had to be allowed. Incidentally, the original 'gamme', or scale of tints for the layer colouring, produced by the War Office was "such a masterpiece of colour printing that no one has been really successful in copying it".[1] The main impediment to progress, however, was the distribution of responsibility among many independent bodies, influenced by national considerations, and the consequent absence of a strong central body to promote uniformity, and to make the published sheets easily procurable. By the outbreak of war in 1939, of the approximate total of 975 sheets required to cover the land surface, 405 had been published, but of these only 232 conformed to the international pattern. It was held by some that there was insufficient material available to map all countries on this scale, and partly for this reason the Geographical Section, General Staff, between 1919 and 1939, produced series of maps such as Africa, 1:2 Million, and Asia 1:4 Million.

If it has not been a complete success, it has had some useful results. The sheet lines have been fairly widely adopted as the framework for national series on larger cales, thus

[1]Hinks, A. R., *Geogr. Journ.*, 94. 1939, 404.

introducing a measure of uniformity in international cartography. The value of the 1/1 M. sheets for the mapping of
distributions on a continental or world-wide scale has also
been recognized. The sheets have been used for example
for the International Map of the Roman Empire, of which
twelve sheets have been published. The recently inaugurated
World Land Use Survey aims at eventually producing maps on
this scale. The value of the latter map in plans for assisting
the under-developed areas would be considerable.

But perhaps the most striking result of the International
Million Map was the impetus it gave to the Million Map of
Hispanic America which was produced by the American
Geographical Society on the initiative of its former Director,
the late Isaiah Bowman. The map follows the specifications of
the International Map quite closely, and in its compilation the
Society has had the approval and assistance of South American
governments. Begun in 1920, it was completed with the publication of the 107th sheet in 1945. The complete map covers an
area whose greatest dimensions are 34×28 feet; a staff of seven
were employed throughout the twenty-five years in research,
compilation and drawing. Compilation required the assessment
of the relative values of a great quantity of survey material,
and research brought to light much useful data preserved in
government and commercial offices. Where survey data were
entirely lacking, written descriptions were used to discover
physiographic characteristics. To indicate the accuracy of the
material employed, a reliability diagram was included in each
sheet, a useful practice since followed by other cartographic
institutes.

The Development of the Atlas

The representation on maps of features other than those
strictly topographical in character was of course not new.
The printed versions of Ptolemy's maps were in reality historical
maps, and most of the great cartographers of the sixteenth and
seventeenth centuries had published maps to illustrate Biblical
history. In the domain of science Edmund Halley had mapped
the tides of the English Channel, and the lines of equal

magnetic declination. His celebrated 'General chart', with its 'Halleyan lines', published in 1701, was an important contribution to the study of terrestrial magnetism.

Some have seen in the maps of John Rocque, with their distinction between arable, pasture and wood, the forerunners of the 'land use' maps of today. At the beginning of the nineteenth century, maps were being used systematically in the new science of geology. William Smith, a pioneer in the science, using fossils to arrange the strata chronologically, had initiated the geological mapping of England and Wales. His 'Delineation of the strata of England and Wales with part of Scotland', in fifteen sheets on a scale of five miles to one inch, was engraved and published by John Cary in 1815, with the geological data hand-coloured; the work was placed on a permanent footing by the establishment of the Geological Survey in 1835, which used the Ordnance Survey One Inch sheets as a base.

It was however the foundation in Germany of geography as a modern study which demonstrated the use of the maps as instruments for specialized research. Both Alexander von Humboldt and Karl Ritter appreciated their value in understanding the distribution and inter-relation of phenomena on the earth's surface, when they advanced the principle of causality as the mainspring of geographical research. Humboldt in particular showed that cartographically a great variety of facts could be represented in an orderly and readily intelligible manner. The voluminous results of his travels and studies in New Spain were accompanied by an 'Atlas géographique et physique', 1812, in which the beginning of this development is apparent. The device of isotherms, or lines of equal temperature is due to him, and he also mapped the areal and altitudinal limits of plants and other phenomena.

His ideas were developed enthusiastically by several disciples, whose work was presented to the general public by the celebrated establishment of Justus Perthes at Gotha. One of these was Adolf Stieler, who, after practical experience in survey, came forward with a somewhat grandiose plan for a general atlas. His specifications included a convenient format; text to accompany each map; the greatest possible accuracy, clarity, and comprehensiveness; uniformity of projection and scale; good paper and printing; careful colouring—and a reasonable price: specifications which still elude the publishers of atlases. The first fascicule of the famous 'Hand-Atlas', under the direction of Justus' son Wilhelm, was issued in 1817. In six years the fifty sheets, as first planned, had appeared, but, almost from the first, supplementary sheets had been added, and what is regarded as the first complete edition, containing seventy plates, came out in 1830. For 100 years, new editions appeared at intervals, down to the publication of a large International Edition in 1930. The atlas was distinguished by its careful engraving and subdued colours applied by hand, later superseded by colour lithography, which gave wider scope to the compilers and reduced costs. Perthes' growing reputation attracted the attention of the man who was endeavouring to put the ideas of Humboldt into execution—Heinrich Berghaus, who had established a school of cartography at Potsdam, and trained notable men like August Petermann. Their co-operation resulted in the important 'Physikalischer Atlas', designed to represent graphically the main phenomena of inorganic and organic nature according to their geographical distribution and divisions, and to bring to life that aspect "which in written studies often lies buried in dead words".

The first edition appeared in 1838, and a second, revised and extended, in 1852. The latter in eight parts and containing in all ninety-four maps, was a remarkable achievement: the contents comprised: meteorology and climatology; hydrography; geology (including a contoured map of Europe at 500 ft., 1,000 ft. and then at intervals of 1,000 ft.); terrestrial magnetism; plant geography (demonstrating the vertical distribution of vegetation, the spread of cultivated plants, etc.); anthropography; and ethnography. Though the data on which the maps were based were understandably often incomplete and defective, and were sometimes interpreted in accordance with theories now abandoned, its comprehensiveness of outlook and general methods have rarely been surpassed, and it set a standard for subsequent work of this kind.

It is pleasant to record, however, that it was not without a rival in Britain, the fruit of the industry and perseverance of

Alexander Keith Johnston. This native of Edinburgh, who had commenced cartography as early as 1830, had, in travel on the Continent, become acquainted with Humboldt and Ritter, receiving encouragement from the former to produce an English equivalent of Berghaus' 'Physikalischer Atlas'. An arrangement was made for the use of Berghaus' material, but this ultimately fell through, and Johnston went to work independently. The statement sometimes made that his 'Physical Atlas' was merely an English edition of the German is incorrect. The Atlas (second edition, 1856) was favourably received, in all some 2,500 copies being sold, and was declared by Ritter to be of higher merit than its German rival. His 'Royal Atlas' of 1859, the German sheets of which were minutely criticized in proof by Prince Albert, was another conscientious piece of work.

The mutual relations between German and British cartographers is further illustrated by the career of August Petermann, who first went to Edinburgh as assistant to Johnston, and later set up as a lithographic printer and map publisher in London in 1847. Here in touch with the Royal Geographical Society he met many of the travellers who were opening up the interiors of the continents. Finally he was induced to join the Perthes firm in Gotha, where his contacts were invaluable in obtaining and publishing the results of contemporary exploration.

Contemporary with Keith Johnston and Petermann were the two John Bartholomews, father and son, who founded the Edinburgh Geographical Institute. They did much to improve the standard of non-official cartography in Britain, and introduced new techniques; John George Bartholomew, the third of the line, for example, initiated the layer-colouring of medium-scale maps. Also noteworthy is J. G. Bartholomew's project of a great 'Physical Atlas' in five volumes, which was to sum up the state of knowledge at the end of the nineteenth century. In 1899 appeared the 'Atlas of Meteorology' designed to be Vol. 3 of the series, and edited by Alexander Buchan. It contained over 400 maps dealing with all the elements of climate and types of weather. This was followed in 1911 by Vol. 5, 'The Atlas of Zoogeography'. No further volumes were produced, but these two, despite later progress in the sciences, are still valuable for reference.

National Atlases

A later development in atlases, arising from the general advance of geography, has been the publication of National Atlases, i.e. atlases dealing with the physical and human factors of a particular country. It is probably not entirely without significance that the earliest of these were produced by young countries or those with a strong national sentiment. The Royal Scottish Geographical Society's Atlas of Scotland, published by J. G. Bartholomew in 1895, is an early example of this type. Though the basis is a layer-coloured map on the scale of two miles to one inch in forty-five sections, it includes smaller-scale maps illustrating the physiography, geology, climate and natural history, prepared by Sir Archibald Geikie and Alexander Buchan. The Atlas of Finland (first edition, 1899; third edition, 1925), an extremely comprehensive and well-produced work, proclaimed its object to be "to assist the people of Finland to know themselves and their country". In addition to physical geography and geology, the second edition covers also hydrology, flora, archæology, and demography (with maps showing the distribution of population by red dots for each ten inhabitants).

Less extensive in scope, and understandably emphasizing economic factors, is the Atlas of Canada (first ed., 1906; second ed., 1915), a more elaborate successor to which was published in 1957. Several European countries, including Czechoslovakia and Germany, followed suit. One of the most comprehensive was the 'Atlas de France' compiled by the French National Committee for Geography, and printed by the Service Géographique de l'Armée. The plates in colour, including maps on scales as large as 1:1 M., analyse the elements of the country and work of the French. Of much wider interest than its title suggests is the 'Atlas of American Agriculture', prepared by O. E. Baker, and published in 1936, after twenty years' work. These clear maps, printed in colour, deal with the elements of "the physical basis including land relief, climate, soils and natural vegetation". The 'Great Soviet

F*

Atlas of the World' is a combination of a general and a national atlas. Volume I (1937) deals with the world and the Soviet Union in general; the second volume (1939) deals in greater detail with the political and administrative units, topography, and economic geography of the Union, some of the 'oblasti' being mapped on a scale of 1:1.5 M. In it, population densities are shown by proportionate circles for the large centres and colour bands elsewhere. The difficult problem of showing the distribution of population effectively has still to be solved satisfactorily. In the 'dot' method, except on large scales, the placing of the dots (representing given units of population) must be to a considerable extent arbitrary. The differential colouring of areas, resembling the layer colouring of topographical maps, is generally preferable, but it is not easy to show sudden and considerable increases in density.

There is no National Atlas of Great Britain to be recorded, despite the strong case made out in 1940 by a committee of the British Association, and supported by the Royal Geographical Society. Material for this exists, however, in the series of maps of Great Britain on the scale of 1:625,000 (about ten miles to one inch) compiled by the Ministry of Town and Country Planning, and published by the Ordnance Survey. These, employing a common base, include such topics as land utilization, types of farming, population density, coal and iron, iron and steel, and economic minerals. The Ordnance Survey have announced the intention of issuing these as 'atlas' maps on half the scale, 1:1.25 M., but so far only the base map has been issued. It is to be hoped that these constituents of a national atlas illustrating "the life, work, wealth, and physical conditions of Great Britain" will not be long delayed.

During the war, attempts were made to induce the public to abandon the 'flat', non-continuous world map of the Mercator type, with the north 'at the top', for others emphasizing the 'round world', and the new relationships inherent in the development of the aeroplane. A striking example was R. E. Harrison's *Fortune* Atlas, 'Look at the world', in which the orthographic projection was used, with unusual orientations. 'Near globes', i.e. geometrical constructions resembling globes which could also be laid out flat, were also devised, an instructive example being E. G. R. Taylor's 'Air age world map' (1945).

This rapid review of modern 'special' atlases has indicated some of the types of maps now being produced. Mention should also be made of others not classifiable as atlases. There are for example the sheets on the scale of one inch to one mile published by the Land Use Survey of Great Britain. This survey was initiated by Professor Dudley Stamp in 1930 "to determine the present use of every acre" of the country. The results are published on the basis of the Ordnance Survey One Inch map in seven colours, differentiating forest and woodland, meadow and permanent grass, arable, heathland, orchards, gardens, etc., and agriculturally unproductive land. The value of such a survey was demonstrated during the war, and it inspired and assisted much similar work in connexion with regional planning. Work based on the International Million map has already been noted: to this may be added the Ordnance Survey Period Maps of Britain, which now range from prehistoric times to the seventeenth century. Also notable are the maps being issued by the American Geographical Society, which will eventually form an 'Atlas of Disease'.

For its representation of relief and its lettering, the map of Europe and the Middle East, produced by the Royal Geographical Society for the British Council, is worthy of study.

It is true to say, therefore, that a new 'reformation of cartography' is now in progress. In each department of map making, from the field to the printing machines, new techniques are being applied. The sphere open to the cartographer has also widened immensely. Efforts have been made, especially during the last war, to develop an understanding among the public of the value and the limitations of the map. The next chapter gives a brief account of progress in cartography since 1953.

REFERENCES

CHEETHAM, G., New medium and small scale maps of the Ordnance Survey. (*Geogr. Journ.*, 107, 1946, 211.)

CLOSE, SIR C. F., The map of England. 1932.

HART, C. A., Air survey. (*R. Geogr. Soc.*), 1948.

HINKS, A. R., Maps and survey. 4th ed. Cambridge, 1942.

—— The science and art of map-making. (*Pres. Address, Section E., Brit. Assocn.* 1925.)

——, Making the British Council Map. (*Geogr. Journ.*, 100, 1942, 123)

HORN, W., Die Geschichte der Gothaer Geographischen Anstalt im Spiegel des Schriftums. (*Petermans Geogr. Mitt.*, 1960, 271–87.)

LEWIS, SIR CLINTON, The making of a map. (R. Geogr. Soc.), 1945.

Ordnance Survey, A description of O.S. small-scale maps (*also* Medium-scale maps, Large-scale plans), Chessington, 1947–49.

PERTHES, B., Justus Perthes in Gotha, 1785–1885, München, 1885.

Royal Scottish Geographical Society, Mapping Britain, Edinburgh, 1960.

WINTERBOTHAM, H. ST. J. L., A key to maps. 1945.

WRIGHT, J. K., Highlights in American cartography, 1939–1949. (*C.R. Congrès internat. de géogr.*, *Lisbonne* 1949. I, 304.)

The Map of Hispanic America on the scale of 1 : 1,000,000. (*Geogr. Review*, 36, 1946, 1–28.)

Cartography in the United States of America. (*World Cartography*, 1, 1951, 81–97.)

CHAPTER XII

CONTEMPORARY CARTOGRAPHY

SINCE the first edition of this book was published, cartography has advanced at a greater pace probably than in any earlier decade. Much thought has been given to the application of contemporary technology to survey, map compilation and printing, and to the style and content of maps and atlases. The value of the map for many purposes—in the development of overseas countries, national planning, commerce and research —is now widely recognized. The technique of survey lies outside the scope of this book, but to indicate progress in this direction, mention may be made of the use of the helicopter for conveying parties rapidly to points difficult of access, of the tellurometer, a radio instrument for the almost instantaneous measurement of distances, and an airborne instrument for recording height profiles of the land surface. In the compilation and printing of maps, the use of plastic sheets has resulted in economies in labour and in costs, while securing precision and clarity. Scribing on plastic sheets and the use of photographic lettering machines has greatly reduced the work of the draughtsman. Such methods have resulted in more varied, and cheaper, maps, thus greatly extending their scope and use.

One particular cartographic problem to which much attention has recently been given is the representation of relief. Traditionally, British atlases have favoured hypsometrical layer colouring, while cartographers of other countries have preferred a background of restricted hachures or hill shading. The colour tints have usually followed the arrangement of the spectrum, from blue for the sea, green for the lowest areas, through yellow for the higher to red and reddish brown for the highest, with the areas above the snow-line in white or bluish-grey. The aim now is to convey a plastic or naturalistic impression of the principal relief features. The line of

progress has been the omission of the contour lines bounding
the layers, and extensive use of oblique hill shading (with the
source of light in the north-west). The colour range has also
been reconsidered, and the principle of 'the higher the lighter'
has been successfully employed in Swiss school atlases. The
effect approaches closely that produced by a relief model, and
is probably as good as can be achieved on small-scale atlas
maps.

There is obviously more scope with topographical maps of
medium and large scale, but care requires to be taken to avoid
a multiplicity of colours which will obscure other detail. (Some
of the considerations have been set out on pp. 157–8 and 161.)
The factor of cost must also enter in, for additional colour
plates are expensive. The most elaborate and striking methods
for areas of high and rugged relief have been evolved in
Switzerland, and the 'Swiss style' with modifications has been
adapted to the requirements of other countries. A characteristic
feature is the finely drawn rock work on precipitous slopes.
Good examples of this style are the maps of the Mont Blanc
region by the Institut Géographique National, Paris, on the
large scale of 1:10,000 with no less than 12 colour printings;
the map of Mount McKinley, Alaska 1 : 50,000 by Bradford
Washburn, executed by the Swiss Federal Mapping Bureau,
though some may feel that here the colours tend to dominate
the map; and the excellent map of Mount Everest by the
German-Austrian Alpine Union. Comparable with these is
the first detailed map of the Mount Everest region, 1 : 100,000,
published by the Royal Geographical Society in 1962. In
regions of less pronounced relief, probably restrained layer
colouring, with additional contours or some form of shadowing
on the layers may offer the best solution.

International Map of the World

With the problems of development facing many new
countries with small cartographic establishments, and without
the funds to embark upon large mapping programmes, renewed
interest is being taken in the International Map of the World
on the scale of 1 : 1 Million (IMW), which can provide a
basis for planning. The liaison work formerly performed by

the Central Bureau at the Ordnance Survey Office, Southampton, has now been transferred to the Cartographic Department of the United Nations, and at Bonn in 1962 an
international conference considered the future of the IMW.
The Map was designed and the specifications agreed upon to
meet rather different conditions than those of the contemporary
world, so that the accepted specifications require amendment
and extension. Methods of simplifying and reducing the costs
of production would also appeal to the new countries unable
to maintain the present high standards, last revised in 1913,
if the present system of making each country responsible for
the sheets covering its territory is continued. It is clear also
that resources could be conserved by co-ordinating work on
the IMW with charts of the International Council of Aeronautical Organizations (ICAO). It has been proposed, for
example, that the sheet lines of the latter might be adapted
for the IMW without great difficulty. The projection (Lambert
Conformal) of the ICAO charts might also be substituted for
the polyconic projection of the IMW. In this case the difference
would not be greater than the changes in the size of the individual sheets due to variations in temperature and humidity of
the atmosphere.

The ICAO World Air Chart itself is but one of a series
designed to meet all the needs of modern air navigation and
transport route planning, navigational aids, approach techniques
and airport facilities. These charts are under constant revision
in the light of technical advances; it is a matter for consideration, for example, whether in an age of jet-propelled aircraft,
flying at great speeds and at great heights, a route planning
chart on the scale of 1 : 2 Million, rather than 1 : 1M, would
not suffice. While on the subject of world communications,
brief reference should be made to developments in hydrographical charts. The use of modern navigational techniques
and echo-sounding apparatus has greatly simplified and improved marine surveys. The Hydrographic Department of the
British Admiralty, the body responsible for the production
and maintenance of accurate hydrographical charts from surveys
by the Royal Naval Surveying Service and many other authorities, has recently introduced further improvements in its charts,

made possible in part by the introduction of colour printing.

Some Map Series

The IMW, in view of its relatively small scale, is essentially a planning map, and for detailed work much larger scales are necessary. For many of the developing countries, particularly in Africa, such maps in the past have been largely provided by, or on the initiative of, European agencies. It is essential that this co-operation should continue. In the case, for example, of Ghana and Nigeria, existing Survey Departments have continued on the same lines after independence, and the links with the Directorate of Overseas Surveys have been maintained. This Directorate, established under the United Kingdom Colonial Development and Welfare Acts, has mapped, mainly by photogrammetric methods, some three-quarters of a million square miles of overseas territories. These topographical maps are mainly on the scale of 1 : 50,000 but the Directorate has also published a number of special maps—geological, soil and land use surveys—in connection with specific development projects. In this connection, mention should also be made of the 'National' atlases produced for African territories, e.g. Ghana and Tanganyika. The latest of these is the excellent 'Atlas of Kenya', 1959, prepared from the national survey and other government sources.

The Ordnance Survey has now replaced the Sixth (New Popular) edition of the One-Inch Map with a Seventh Series, the complete 190 sheets being issued between 1952 and 1960. The whole of Great Britain is now, for the first time, covered by one uniform series. Relief is still represented by contours only; the map has been entirely re-drawn and some minor changes in style introduced, e.g. the black symbols of types of woodland have been restored. The term 'Series' is used to denote that the individual sheets will be revised as necessary, but to cope with new developments the road information will be revised whenever a sheet is reprinted.

Though the Ordnance Survey has abandoned hill-shading and layer colouring on the One-Inch Series, these have been retained on other scales. They are employed on the One-Inch Tourist Maps of areas of special interest (cf. The North York

Moors, 1958) and on the new Quarter-Inch Map (actually the scale, slightly larger, is 1 : 250,000 but the name is being temporarily retained). A striking example of the latter is the Western Highlands sheet.

General atlases

In Britain, makers of atlases have been particularly active in this period. A major work was 'The Times Atlas of the World: Mid-Century edition', edited by Dr. John Bartholomew, whose firm had produced an earlier Times Atlas in 1922. In five volumes (1955–59) this work, using the familiar layer colouring, provides unusually large scale coverage for the world and a very comprehensive index, divided among the five volumes. This is a work essentially for the reference library, and several new atlases for general use also appeared. One of these may be mentioned here, since its publication signalized the advent of a new cartographic publisher—The Oxford Atlas of the World, by the Cartographic Department of the Clarendon Press (1st ed. 1951).

Notable among the many continental atlases is the 'Atlante Internazionale' of the Touring Club Italiano, an international edition of which was published in 1955–56 to mark the sixtieth anniversary of the Club. This is a good clear reference atlas, which states the sources from which the sheets have been compiled and adheres closely to the local official spelling for place names. This standard is well maintained by G. Dianelli's 'Atlante fisicoeconomico d'Italia' also published by the T.C.I.

Among new German atlases, 'Bertelmanns Grosser Welt-atlas', 1961, carefully designed and executed, takes a high place. The Soviet output of general atlases is significant. The 'Atlas Mira' (Atlas of the World), 1954, is comparable with the best contemporary atlases. The great 'Morskoi Atlas' (Atlas of the seas), 2 vols. and index, 1950–53, though primarily oceanographical, is sufficiently comprehensive to be included here. In addition to bathymetrical maps, in colour with soundings, of the oceans and seas (some on scales as large as 1 : 1·5M), currents and winds, it includes layer coloured relief maps of the land surface, climatic and magnetic maps.

National and special Atlases and maps

It is not possible to enumerate here all the National Atlases that have been published since 1953, but the following deserve attention. A provisional edition of the ambitious 'National Atlas of India', edited by Professor S. P. Chatterjee, appeared in 1959.

A much enlarged edition of the 'Atlas of Canada' was published in 1957 (see p. 169), and another Dominion is represented by the 'Atlas of Australian Resources', 1954–60, with maps generally on the scale of 1 : 6M. One of the earliest national atlases, that of Finland, reached its fourth edition in 1961; other European countries to have produced similar works are Sweden and Belgium. The Ordnance Survey has published further sheets in its Ten-Miles to the Inch Series, and a comprehensive 'Atlas of Britain', with maps mostly on the scale of 1 : 2M, is to be published by the Clarendon Press. It is probably only a matter of time before every developed country possesses a national atlas, and a Commission of the International Geographical Union is studying the standardization of such works, and the possibility of applying their methods to problems of world-wide extent.

The Cartographic Department of the Clarendon Press has published several special atlases. Notable is the series of regional economic atlases, prepared with the assistance of the Economist Intelligence Unit, of which two volumes have appeared, for the Middle East and the U.S.S.R. Given the limitation of scale and the difficulties of securing and presenting the data, these are useful reference works. There are obvious difficulties in deciding, for example, where to place a dot representing production of 50,000 tons of wheat.

An unusual special atlas is the 'Relief form atlas' (in French and English editions) published by the Institut Géographique National, Paris, in 1956. This covers systematically all types of landforms; plates can be viewed stereoscopically, and are accompanied by relevant air photographs. The Institut has also developed a method of producing accurate relief models from plastic sheets, and it is instructive to compare these with the relevant topographical sheets. In this way the problems confronting the cartographer and the methods he has used to solve them can be appreciated.

In Soviet Russia the technique for special mapping has been highly developed. Even before the Revolution a number of economic atlases had been published. Lenin was quick to perceive their value for "increasing and developing the productive forces of the State," and, incidentally, for propaganda. Geographical education is organized very largely as an instrument for building-up and strengthening the Lenin-Marxist state. All graduates are obliged therefore to spend a certain period in prescribed research, so that there is a corps of specialized workers continuously employed as members of expeditions in the field or in research institutes, gathering and working-up a vast amount of material for the preparation of special maps which contribute to the preparation of Five Year Plans and to a detailed knowledge of the geography of the U.S.S.R. Much of this work finds its way eventually, on greatly reduced scales, into special atlases. The aim is as far as possible to present dynamic trends rather than merely to record existing conditions. By a lavish use of colour and a great variety of symbols, drawn proportionately, the size, characteristics, structure and flow of industry is thus presented visually. Soviet economic mapping is distinguished rather by its intensity and its wide scope than by the evolution of striking new principles. An important publication is 'The Agricultural Atlas of the U.S.S.R.' (Moscow: Directorate of Geodesy and Cartography, 1960), with maps for the separate republics.

The value of larger scale regional atlases has also been appreciated. Here the standard was set by the 'Atlas of the Leningrad and Karelian region', 1934. A more recent example is the 'Atlas of the White Russian S.S.R.', 1958. Among special maps are a tectonic map of the U.S.S.R. (1 : 5M) and a notable forest map (1 : 2·5M).

Epilogue

What is the future of maps and atlases likely to be? It is always rash to attempt a prophecy, but of one thing we may be tolerably certain. The demand for maps will continue to grow rapidly.

Some indication has already been given of the variety of special maps which are being produced to meet new needs, and these will increase with profounder knowledge of the world around us. But while special maps dealing with a specific topic can be multiplied almost indefinitely, the need is being realized for maps which will present the inter-relation of a number of distributions. A simple example is the correlation between soils and crops; to this might be added rainfall, the accumulated monthly temperatures, and even the distribution of population (since this may influence, for example, the location of market gardening). How to present all this data accurately and clearly is a matter for close co-operation between the scientists involved and the cartographer. The character of the data and the relative sizes of the categories to be plotted will affect the scale (perhaps also the projection, if the area is considerable) the types of conventional signs, the number of colours and their range, and other details. These can only be settled by discussion between the scientist and the cartographer, and it is the duty of the former to explain the purpose of the map. Unless this is perfectly understood at the start, the map will be a failure. But it should be understood that a map cannot of itself solve a problem. It can demonstrate significant correlations, or their absence, and in this way satisfy the scientist that he is proceeding on the right lines, or suggest fruitful subjects for further research.

The range of subjects covered by maps will therefore increase, and this will be accompanied by growing complexity of the data shown. In view of this it may be anticipated that all maps will come to bear a precise statement of their purpose and the sources on which they are based—a statement resembling the 'index of reliability' which now accompanies some topographical series. Is it too much to hope, in this respect, that all maps will bear at least the date of publication? The degree and the rate at which such maps will circulate more widely will in part depend upon developments in production and printing techniques, leading to the wider use of colour and to a reduction in costs.

In regard to the topographical map, where the representation of the relief is a major feature, the scope for change is not so great. Hachuring or hill-shading alone, now on the decline, is likely to disappear entirely. There is room for further development in portraying relief 'plastically', though it is to be hoped that contour lines will not be abandoned completely. It has been asked whether '3D', or relief, maps will become more general. They certainly have their uses particularly in education, when features can be emphasized by the exaggeration of the vertical scale, but as a general substitute for sheet maps they are inconvenient, and are no more expressive than a carefully designed topographical sheet.

As for the 'general' map, now largely used as a means for getting from one place to another, it would appear that they will generally be on smaller scales. Just as the One-Inch map served the 'foot-slogger', and the early motorists used the Quarter-Inch, so the convenient scale for the traveller of the future will decrease. There will, however, always be room for the specially designed larger scale maps of particular areas for the walker and the mountaineer, not to mention the pot-holer.

Much of what has been said above will apply to the atlas of the future. The general atlas, principally for locating places and routes, will share in the improvement in style, with a tendency to less crowding-in of names. A good comprehensive index should allow the quick location of localities of lesser importance not included on the maps. With atlases which aim at supplying more, and more varied, information, the line of progress is less clear. Most large atlases already include a section of special or topical maps.

As the more specialized 'national' atlases will continue to increase in number and in range of content, one might hazard the guess that the general atlas would evolve on parallel lines into a supra-national, or world, atlas. However, this would probably involve questions of finance: but, just as the sixteenth-century collector had his own selection of maps bound into an atlas, is it Utopian to foresee a map publisher offering a minimum number of map plates from a comprehensive list in a spring-binder to prospective clients?

Whatever the future may hold, it is at least clear that the problems and opportunities facing the cartographer will not decrease.

REFERENCES

Modern Cartography. Basic maps for world needs. Especially Pt. II: study on modern cartographic methods. United Nations, 1949.

Comptes-rendus des Séances de la 2 me Conférence Internationale, Paris, 1913 (carte internat. du monde au millionième)

The International Map of the World on the millionth scale and the international·co-operation in the field of cartography (*World Cartography*), 3, 1953, 1–12.)

BALCHIN, W. G. V., Atlases today. (*Geogr. Mag.*, 32, 1959–60, 554.)

CRONE, G. R., The future of the International Million Map of the World. (*Geogr. Journal*, 128, 1962, 36)

GARDINER, R. A., A re-appraisal of the International Map of the World (IMW) on the millionth scale. (*Internat. Yb. of Cartography*, 1, 1961. 31–50.)

SHERMAN, J. C., New horizons on cartography; functions, automation, and presentation. (*Internat. Yb. of Cartography*, 1, 1961, 13.)

NOTE

The 'Atlas of Great Britain and Northern Ireland', Clarendon Press, 1963, is in effect a national atlas of Britain. Other additions to this category are the 'Atlas of the Union of South Africa', Pretoria, 1960, and the 'Atlas of Uganda', Kampala, 1962. The Royal Geographical Society has published the 'National Atlas of Disease Mortality in the United Kingdom', in co-operation with the Medical Research Council, 1963. A Second Land Utilization Survey of England and Wales, mapping land use on the scale of $2\frac{1}{2}$ inches to the mile, has been initiated by Miss Alice Coleman, and several sheets have been published.

APPENDIX I

GENERAL WORKS ON CARTOGRAPHY

These books, for the most part not quoted in the chapter references, deal with various aspects of the making, history and use of maps. Many of the volumes of reproductions listed after them contain explanatory texts and bibliographies.

BAGROW, L., Geschichte der Kartographie. *Berlin*. 1951.

CHAMBERLIN, W., The round earth on flat paper. (Nat. Geogr. Soc.) *Washington*. 1947.

DESTOMBES, M., Catalogue des cartes gravées au XVe siècle. (Union Géogr. Internat.: Rapport, Commission pour la bibliographie des cartes anciennes.) *Paris*. 1952.

FISHER, I. and MILLER, O. M., World maps and globes. *New York*. 1944.

Geographical Journal. 1893 onwards. Quarterly.

HINKS, A. R., Maps and survey. 5th Ed. *Cambridge*. 1947.

Imago Mundi. 1935 onwards. Annual.

International Yearbook of Cartography. 1961 onwards.

KIMBLE, G. H., Geography in the middle ages. 1938.

LYNAM, E., The mapmaker's art. 1953.

MONKHOUSE, F. J. and WILKINSON, H. R., Maps and diagrams; their compilation and construction. 1952.

RAISZ, E., General cartography. 2nd ed. *New York*. 1948.

ROBINSON, A. H., The look of maps. 1952.

SKELTON, R. A., Decorative printed maps of the fifteenth to eighteenth centuries. 1952.

Surveying and Mapping. 1940 onwards. Quarterly. *Washington*.

TOOLEY, R. V., Maps and map makers. 1949.

APPENDIX II

REPRODUCTIONS OF EARLY MAPS AND CHARTS

ALMAGIÁ, R., Monumenta italiae cartographica: reproduzioni di carte generali e regionali d'Italia dal secolo XIV al XVII. *Firenze.* 1929.

Anecdota cartographica septentrionalia; ed. A. A. Björnbo and C. S. Petersen. *Haunia,* 1908. (H. Martellus' map of Scandinavia, etc.)

BAGROW, L., Anecdota cartographica. 1–4. *Stockholm.* 1935–51. (Maps of Ukraine, etc.)

——, The 'Atlas of Siberia' of Semyon U. Remezov. (*Imago Mundi* Suppl.) The Hague, 1959.

Bibliothèque Nationale, Choix de documents géographiques conservés à la. *Paris.* 1883. (Carte pisane, Catalan atlas, c. 1375.)

BRITISH MUSEUM, (i) Map of the world designed by G. M. Contarini, 1506. 1924. (ii) Sir F. Drake's voyage, 1577–80; two contemporary maps. 1927. (iii) Four maps of Gt. Britain by Matthew Paris, c. 1250. 1928. (iv) Six early printed maps, 1928. (v) Maps of C. Saxton, engraved 1574–9. 1936–9.

CARACI, G., Tabulae geographicae vetustiores in Italia adservatae; Reproductions of Ms. and rare printed maps of the great discoveries, 3 pts. *Firenze.* 1926–32.

COOTE, C. H., Facsimiles of three mappemondes, 1536–50. (Biblioteca Lindesiana), 1898. (Maps of Dieppe School.)

CORTESÃO, A. Z., and A. TEIXEIRA DA MOTA, Portugaliae Monumenta Cartographica. 4 vols. Lisbon, 1960. (A comprehensive series of Portuguese charts to the seventeenth century, many in colour.)

DESTOMBES, M., La mappemonde de Petrus Plancius. *Hanoi.* 1944.

FISCHER, J., Claudii Ptolemaei Geographiae Codex Urbinas graecus 82. Leipzig, 1932. (Also reprodns. of other Ptolemy MSS.)

—— and VON WIESER, F., Die älteste Karte mit dem Namen Amerika, 1507, und die Carta marina, 1516, des M.Waldseemüller. *Innsbruck.* 1903.

FITE, E. D. and FREEMAN, A., A book of old maps delineating American history ... to close of Revolutionary war. *Cambridge, Mass.* 1926.

FOCKEMA ANDREAE, S. J., and B. van't HOFF, Christiaan Sgroten's Kaarten van de Nederlanden (Maps of the Netherlands). Leiden, 1961.

HANTSZCH, V. and SCHMIDT, L., Kartographische Denkmäler zur Entdeckungsgeschichte von Amerika, Asien, Australien und Afrika. *Leipzig.* 1902. (N. Desliens' world map, 1541; D. Homem, Atlante maritimo, 1568.)

JOMARD, M., Les monuments de la géographie, ou recueil d'anciennes cartes européennes et orientales. *Paris.* 1842-62. (La Cosa, Mercator, etc.)

KRETSCHMER, K., Die Entdeckung Amerika's in ihrer Bedeutung für die Geschichte des Weltbildes. Altas. *Berlin.* 1892.

KUNSTMANN, F., Atlas zur Entdeckungsgeschichte Amerikas. *Munich.* 1859.

LEPORACE, T. Gasparrini, and R. ALMAGIÀ, Il mappamondo di Fra Mauro. Venice, 1956. (A full-size reproduction in colour, with introduction, notes and identifications of place-names.)

MARCEL, G., Reproductions de cartes et de globes relatifs à la découverte de l'Amérique du XVIe au XVIIIe siècle. *Paris.* 1893.

—— Choix des cartes et de mappemondes des XIVe et XVe siècles. *Paris.* 1896. (A. Dulcert, Mecia de Villadestes, Soleri.)

MERCATOR, G., Drei Karten von G. Mercator; Europa-Britische Inseln-Weltkarte. *Berlin.* 1891.

Monumenta cartographica vaticana. I. Planisferi, carte nautiche e affini dal secolo XIV al XVII. 1944. II. Carte geogr. a stampa di particolare pregio o rarità dei secoli XVI e XVII. *Vatican City.* 1948. Ed. by R. Almagià.

MORTENSEN, H., and A. LANG, Die Karten Deutscher Länder im Brusseler Atlas des Christian s'Groten, 1573. Gottingen, 1959.

NORDENSKIÖLD, A. E., Facsimile atlas to the early history of cartography. *Stockholm.* 1889. (Ptolemy (Rome, 1490), Berlinghieri, Apian, Ortelius, Mercator.)

—— Periplus; an essay on the early history of charts and sailing directions. *Stockholm.* 1897. (P. Vesconte, Ribero, Gastaldi.)

NØRLUND, N. E., Danmarks Kortlaegning; Johannes Mejers Kort over det Danske Rige (and other publications). Geodaetisk Institut, København, 1942–4.

ONGANIA, F., Raccolta di mappamondi e carte nautiche del XIII al XVI secolo. 15 pts. *Venice.* 1875–81. (Vesconti, Bianco, Genoese map of 1457, B. Agnese.)

Remarkable maps of the fifteenth, sixteenth and seventeenth centuries reproduced in their original size. *Amsterdam: F. Muller,* 1894–8. (Pts. ii, iii, seventeenth cent. Dutch maps of Australia.)

ROYAL GEOGRAPHICAL SOCIETY, (i) Map of the world by J. Hondius, 1608. Mem. by E. Heawood, 1927. (ii) English county maps. Mem. by E. Heawood. (iii) Portolan chart of A. de Dalorto. Mem. by A. R. Hinks, 1929. (iv) Catalan world map. R. Bibl. Estense at Modena. Mem. by G. H. Kimble, 1934. (v) The world map ... in Hereford Cathedral, c.1285. Mem. by G. R. Crone, 1953. (vi) The Map of Great Britain, c. A.D. 1360, known as the Gough Map. Ed. by E. J. S. Parsons, and publ. in co-operation with the Bodleian Library, Oxford, 1958. (vii) Early maps of the British Isles A.D. 1000–A.D. 1579, with a memoir and notes by G. R. Crone. 1961.

SANTAREM, VCTE DE, Atlas composé de mappemondes, de portulans et de cartes hydrographiques et historiques depuis le VIe jusqu'au XVIIe siècle. *Paris*. 1842–53. (Fra Mauro; early charts of W. Africa.)

STEVENSON, E. L., Maps illustrating early discovery and exploration in America, 1502–30. *New Brunswick*. 1906.

—— Marine world chart of N. de Canerio, 1502. *New York*. 1908.

WIEDER, F. C., Monumenta cartographica. 5 vols. *The Hague*. 1925–33. (Plancius, Blaeu, Vingboons.)

YUSSUF KAMAL, PRINCE: Monumenta cartographica Africae et Aegypti, 14 vols. 1926–39. (Reproduces all maps and charts down to the age of discoveries which include any part of Africa.)

NOTE ON MAP COLLECTIONS

The student wishing to pursue the study of the history of cartography will find much material in the Map Room and the Department of Manuscripts of the British Museum. In addition to early manuscripts, such as those of Matthew Paris, numerous early marine charts and Tudor MS maps, the work of General Roy in the Highlands of Scotland and the early maps of the Ordnance Survey can be consulted. Many engraved atlases and maps are unique. The R.G.S. has a representative collection of atlases from the 15th century, and MS maps of 19th-century explorers. The Bodleian Library, Oxford, also has numerous MS and rare printed maps, including the collection bequeathed by the antiquarian and topographer, Richard Gough. The Bibliothèque Nationale, Paris, has an unrivalled collection of MS and printed maps. It is particularly rich in early marine charts, and has the great MS Catalan Atlas, the world map of Sebastian Cabot, and the d'Anville collection.

INDEX

Dates are given for the more important cosmographers, cartographers, and surveyors.

188

DATE DUE

MY 4 '68		
RESERVE		
RESERVE		
AP 22 '69		
GAYLORD		PRINTED IN U.S.A.